COLORING ATLAS
OF HUMAN ANATOMY
Second Edition

COLORING ATLAS
OF HUMAN ANATOMY
Second Edition

Stephen W. Langjahr and Robert D. Brister
Antelope Valley College

THE BENJAMIN/CUMMINGS PUBLISHING COMPANY. INC.

Redwood City, California ■ Menlo Park, California ■ Reading, Massachusetts ■ New York
Wokingham, U.K. ■ Don Mills, Ontario ■ Amsterdam ■ Sydney ■ Singapore ■ Tokyo
Bonn ■ San Juan ■ Madrid

Sponsoring Editor: Melinda Adams
Assistant Editor: Diane Honigberg
Production Editor: Anne Friedman
Production Assistant: Andrew Marinkovich
Senior Systems Consultant: Guy Mills
Editorial Assistant: Sami Iwata
Cover Designer: Mark Ong
Cover Photographer: Ted Kurihara
Text Designer/Production: Irene Imfeld

11 12 13 14 15 16 17 18 19 20-CRS-05 04 03 02 01 00

ISBN 0-8053-4020-3

The Benjamin/Cummings Publishing Company, Inc.
390 Bridge Parkway
Redwood City, California 94065

PREFACE TO THE INSTRUCTOR

The second edition of the *Coloring Atlas of Human Anatomy* has been carefully revised to make it more versatile and even easier to use. Some of the unique features of this edition include:

■ Nearly 250 finely detailed line drawings are designed to ensure clear labeling and color-coding.

■ Uncluttered pages are easy-to-use with task icons and consistent page design.

■ Comprehensive answers at the end of the book allow students to check their work quickly.

■ Extensive coverage of skeletal muscle anatomy, with an emphasis on derivation of concise origins, insertions, and actions, including a summary table of muscle origins, insertions, and actions.

■ Optional Hypercard™ tutorial software corresponds with the anatomical illustrations in the book and reviews basic terminology, poses questions, and verifies student responses.

Application

This review of human anatomy combines many of the best features of coloring workbooks and study guides. It will be most helpful for the laboratory component of undergraduate courses in human anatomy, whether taught separately or in combination with physiology. It is not designed to accompany any particular text or laboratory manual, although it uses chapter headings and artwork that are often used in the discipline. The *Coloring Atlas* can be used by itself or in tandem with the accompanying software.

Design Concept

The *Coloring Atlas of Human Anatomy* is intended to supplement laboratory and lecture instruction. Students should have some knowledge of anatomy, through study and dissection, in order to use the book most effectively. The *Coloring Atlas* is designed to help students derive, rather than simply memorize, basic anatomical information.

Presentation

The book is divided into fifteen modules extending through the body systems in a traditional sequence. Emphasis is given to material normally included in introductory anatomy laboratories where consideration of structure supersedes that of function. When the student has completed each module, he or she will have a handy and personalized laboratory guide for further reference and review.

Students using the *Coloring Atlas of Human Anatomy* are directed to label and/or color-code the anatomical drawings, as well as summarize additional information and definitions in tabular form. Enhanced learning and accuracy are insured through the comprehensive answers provided at the end of the book.

Tutorial Software (Optional)

The *Coloring Atlas* also has a software tutorial component that corresponds to the figures in the book. Although the *Human Anatomy Coloring Atlas* can be used without the accompanying tutorial software, when coupled with a computer it becomes an interactive, self-paced tutorial reviewing basic terminology and anatomical relationships through provocative questions and discussion of student responses.

The computer text files will run on any Macintosh computer outfitted with Hypercard™, Version 1.2 or newer. The program is easy to use because there are no program commands or detailed procedures to remember. Simple instructions appear continually on the screen to direct the user through the exercises.

Each page of the *Coloring Atlas* features a computer icon in the lower outside corner that shows the computer questions corresponding with the artwork on that page. This allows the user to skip around the book and easily find computer questions relating to any illustration.

It is not possible to use the software without the book because the questions contained on the disk refer to specific figures in the book for visual orientation. Nonetheless, it is possible to use the book without the software.

Ordering the Software

To obtain a copy of the tutorial software for your class, contact your local Benjamin/Cummings Sales Representative or write to:

Connie Cesarin
Sales and Marketing Systems
Benjamin/Cummings Publishing Company
390 Bridge Parkway
Redwood City, CA 94065

Ask for the Langjahr/Brister, second edition tutorial software (ISBN #34021-1). The program is not copyright protected, so copies can be readily distributed to your students.

Acknowledgments

This book has involved the specialized talents of many people.

The use of the illustrations for coloring purposes has been made possible by the superb drawings created by Fran Milner and Sibyl Graber-Gerig, who are illustrators for other outstanding Benjamin/Cummings anatomy and physiology textbooks. Supplemental art was rendered by Tamiko Murakami, Technical Illustrator for Antelope Valley College. Donna Dinger transcribed all of our text for both editions, and she helped us catch numerous errors and inconsistencies.

The second edition of the software, intended for use with the Macintosh family of computers, has been designed to our specifications by Tom Dallman and Guy Mills at Benjamin/Cummings. We are especially grateful to Guy Mills for helping to create a trouble-free, easy-to-use program.

We have enjoyed a most cordial and confident spirit at Benjamin/Cummings. The merits of a dual modality learning tool were steadfastly supported by our first editor, Andy Crowley. Major changes in the manual's format were inspired by our new sponsoring editor, Melinda Adams, and her support and enthusiasm have made this revision possible. The meticulous contributions of Diane Honigberg and Anne Friedman have been particularly important in bringing this much improved edition to completion, especially in the areas of pedagogical style and consistency.

To students or instructors who will be using the *Coloring Atlas of Human Anatomy* and its software, we welcome your comments and invite you to contact us directly with suggestions or concerns that may arise. The creation of this atlas was both fun and challenging for the authors. We hope that instructors and students use it with corresponding results.

Stephen W. Langjahr
Robert D. Brister
Antelope Valley College
3041 West Avenue K
Lancaster, California 93536

PREFACE TO THE STUDENT

How to Use this Book

The *Coloring Atlas of Human Anatomy* is an excellent study tool to accompany your human anatomy or human anatomy and physiology course. Learning is enlivened by a variety of visual and conceptual exercises. The illustrations on each page can be colored and labeled. Tables are provided to organize definitions or summarize information. In lieu of wordy instructions, task icons direct your activity. The meaning of each icon is given below.

Look for the **Identify Icon** near the **upper left corner** of most pages; there will be a list of anatomical structures below it that need to be identified on that page's artwork. The magnifying glass instructs you to examine the art on the page to locate and identify the structures indicated by the terms listed. Structures that should be identified have blank lines next to them for you to fill-in. Identify answers, found in the back of the book, are referenced by uppercase letters.

All muscle actions (Hint: Abduction, Adduction, Depression, Elevation, etc.)

In a few instances, all the identify terms are not listed under the magnifying glass. Instead, a **"hint"** is included, so you are challenged to think of the names of the structures as well as to identify and label them.

The **Coloring Icon** appears along the **left margin** of most pages, with a list of anatomical structures next to it. The crayon directs you to locate and color the listed structures, wherever they appear on the page. You should then fill-in their outlined names with the same color for quick visual cross-referencing. All structures that need to be colored are coded on the figures with numbers inside black circles. These numbers also correspond with the answers section.

The **Definition Icon** is located in the **right margin** of the page. It directs you to provide a short definition for a given term or information about a concept. It is used in conjunction with tables or lists associated with illustrations. Answers in the back of the book are referenced with lowercase letters.

The **combination of Identify and Coloring Icons** is used when you need to identify, label, and color all the listed structures on a page.

For best results, we recommend that you color the figures with **colored pencils** because they do not hide detail, smear, or bleed through the page. Please equip yourself with about a dozen different colors, and then test the coloring tools that you plan to use on the following blank page in order to make sure the color does not bleed through the paper.

How to Use the Book with a Macintosh Computer (optional)

Although the *Human Anatomy Coloring Atlas* can be used without a computer, it becomes a broader, more interactive tutorial if you choose to use the available software. Referring to specific figures in the *Human Anatomy Coloring Atlas,* the software poses questions and provides explanations beyond the scope of the coloring and identification exercises.

Each page of the *Coloring Atlas* features a computer icon in the lower outside corner, which shows the computer questions that correspond with the figures on that page. This allows the user to skip around the book and easily find questions relating to any illustration. It is not possible to use the software without the book because the questions contained on the disk refer to specific figures in the book for visual orientation. Nonetheless, it is possible to use the book without the software.

The computer text files will run on any Macintosh computer outfitted with Hypercard™ , Version 1.2 or newer. The program is easy to use because there are no program commands or detailed procedures to remember. Simple instructions appear continually on the screen to direct you through the exercises. You can color the illustrations as you move through the questions or afterwards. Either way, your *Coloring Atlas* becomes a useful study aid to help you prepare for examinations, once you have completed the coloring and labeling of a module.

Obtaining the Software

To obtain a copy of the tutorial software to accompany the *Human Anatomy Coloring Atlas,* ask your instructor to contact a Benjamin/Cummings Sales Representative. The software is not copyright protected and may be readily shared with others.

COLOR TESTING PAGE

CONTENTS

Preface to the Instructor v

Preface to the Student ix

Module 1 **Anatomical Terminology** 1

Module 2 **Histology** 7

Module 3 **The Skeleton—Bones and Surface Markings** 19

Module 4 **Arthrology** 47

Module 5 **Muscle Actions** 55

Module 6 **Skeletal Muscles** 59

Module 7 **The Circulatory System** 121

Module 8 **The Central Nervous System** 143

Module 9 **The Peripheral Nervous System** 153

Module 10 **The Special Senses** 169

Module 11 **The Endocrine System** 175

Module 12 **The Respiratory System** 183

Module 13 **The Digestive System** 187

Module 14 **The Urinary System** 197

Module 15 **The Reproductive System** 203

Answer Key 209

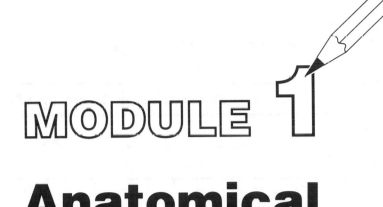

MODULE 1

Anatomical Terminology

■ **Table 1.1 Common Anatomical Terminology**

Directional Term	Definition or Common Meaning
Anterior (ventral)	
Posterior (dorsal)	
Superior (cranial)	
Inferior (caudal)	
Medial	
Lateral	
Proximal	
Distal	
Prone	
Supine	
Superficial	
Deep	

Regional Term	
Dorsum	
Plantar	
Palmar	
Groin (inguinal)	
Axilla	

IDENTIFY

**Axilla
Dorsum
Distal
Groin
Inferior
Lateral
Medial
Palmar
Proximal
Superior**

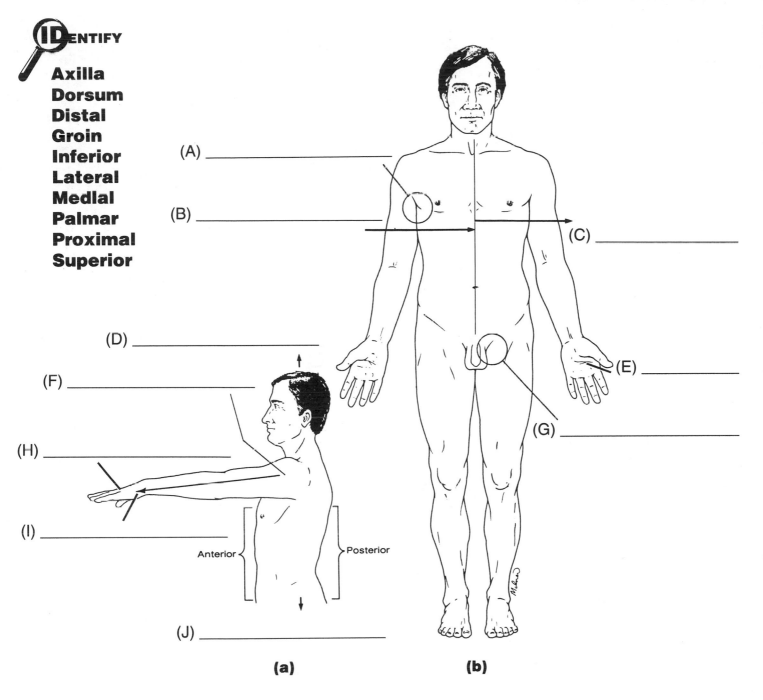

(A) _____

(B) _____

(C) _____

(D) _____

(F) _____

(H) _____

(I) _____

(E) _____

(G) _____

Anterior   Posterior

(J) _____

(a)   (b)

■ **Figure 1.1 Directional and regional terms illustrated,
(a) lateral and (b) anterior perspectives**

Standard Anatomical Position:

_____

_____

_____

Q1-11

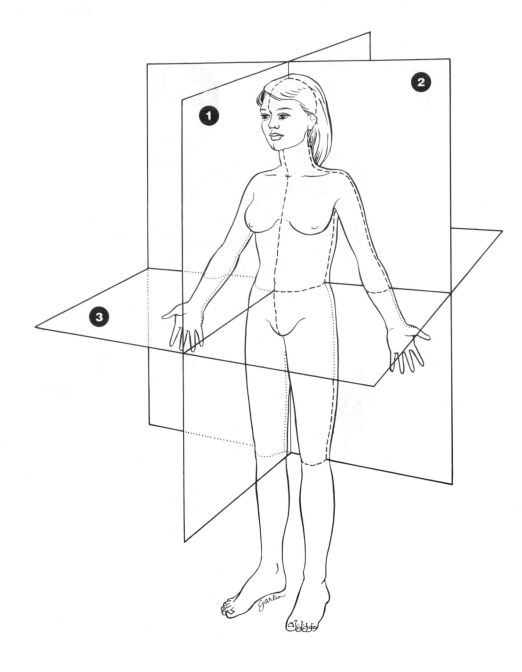

■ **Figure 1.2 Body planes**

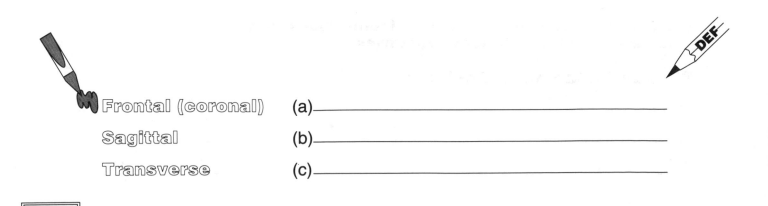

Frontal (coronal)   (a)—————————————————

Sagittal   (b)—————————————————

Transverse   (c)—————————————————

Q12-16

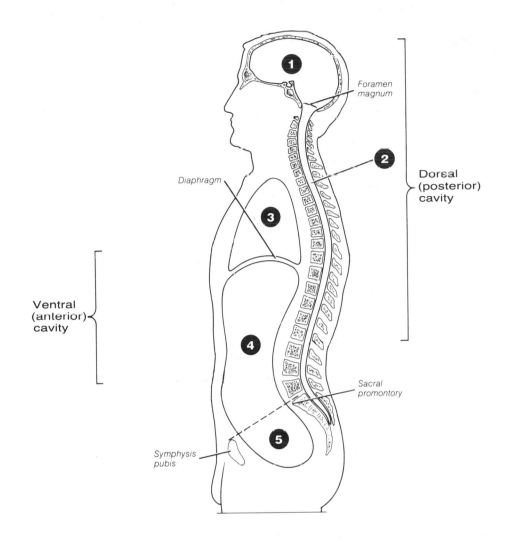

Foramen magnum

Dorsal (posterior) cavity

Diaphragm

Ventral (anterior) cavity

Sacral promontory

Symphysis pubis

■ Figure 1.3 Major body cavities, shown in _____ section

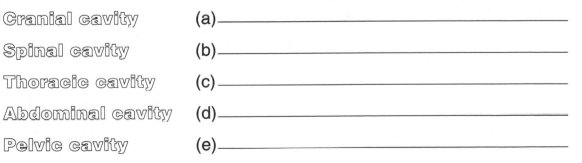

Name	List main visceral components
Cranial cavity	(a)_____
Spinal cavity	(b)_____
Thoracic cavity	(c)_____
Abdominal cavity	(d)_____
Pelvic cavity	(e)_____

Q17-25

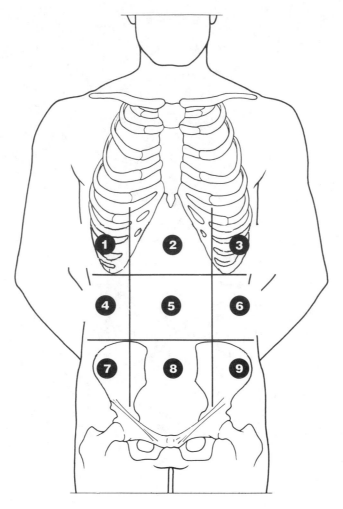

■ **Figure 1.4 Abdominal regions**

Epigastric

Hypogastric

Left hypochondriac

Left iliac

Left lumbar

Right hypochondriac

Right iliac

Right lumbar

Umbilical

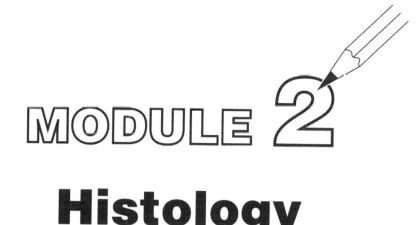

MODULE 2

Histology

IDENTIFY

For each figure in Module 2, identify the tissue depicted. Space is provided below each figure. Illustrated tissues are listed below:

Connective Tissue
Adipose tissue
Areolar connective tissue
Bone
Dense irregular connective tissue
Dense regular connective tissue
Elastic cartilage
Elastic connective tissue
Fibrocartilage
Hyaline cartilage

Epithelial Tissue
Pseudostratified ciliated
  columnar epithelium
Simple cuboidal epithelium
Simple columnar epithelium
Simple squamous epithelium
Stratified columnar epithelium
Stratified squamous epithelium
Transitional epithelium

Muscle Tissue
Cardiac muscle
Skeletal muscle
Smooth muscle

Nervous Tissue
Multipolar neurons
Spinal cord, section

Throughout Module 2, you can use colors typically seen in prepared slides or text photomicrographs. If histology photomicrographs are not available, then the following general guidelines may be applied:

Cilia ordinarily stain the same color as cytoplasm, which is pink.
Collagenous fibers are usually stained pink.
Connective tissues associated with epithelia often have purple nuclei with a pink matrix.
Cytoplasm is ordinarily pink; cytoplasm of cardiac and skeletal muscle cells has alternating pink and white striations.
Elastic fibers are most commonly black.
Ground substance of connective tissue frequently appears light blue.
Matrix of cartilage is light blue around the cells and light pink elsewhere.
Nuclei usually stain purple.
Reticular fibers generally appear black.

■ Figures 2.1–2.4: Epithelial tissues

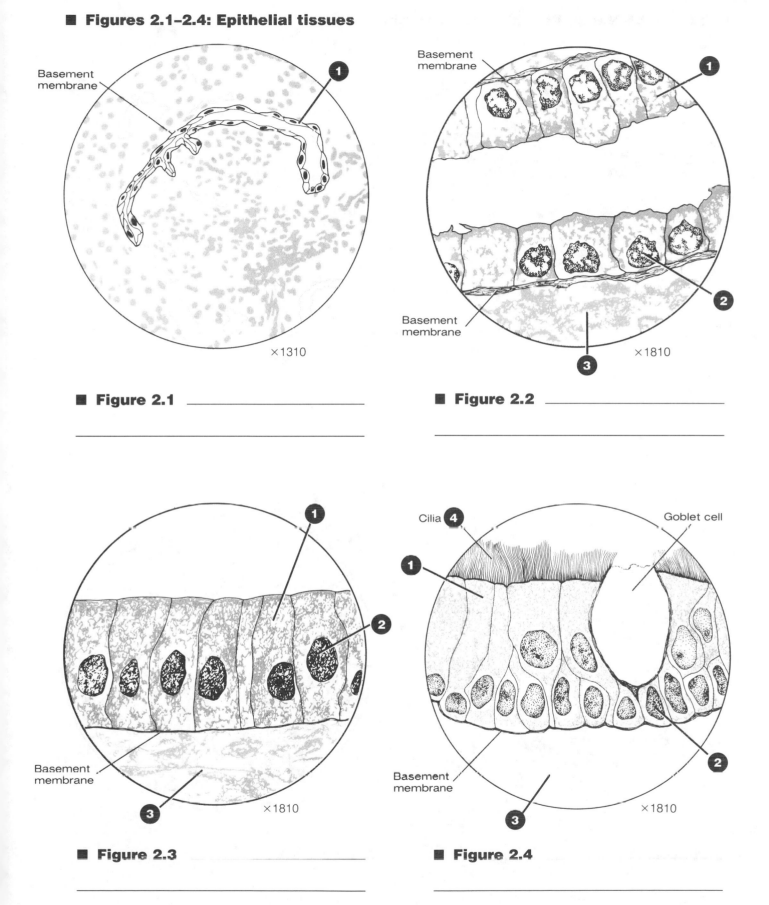

Basement membrane

1

×1310

■ Figure 2.1 _____

_____

Basement membrane

1

2

Basement membrane

3

×1810

■ Figure 2.2 _____

_____

1

2

Basement membrane

3

×1810

■ Figure 2.3 _____

_____

Cilia 4

Goblet cell

1

2

Basement membrane

3

×1810

■ Figure 2.4 _____

_____

Q1-5

■ **Figures 2.5–2.8: Epithelial tissues, continued**

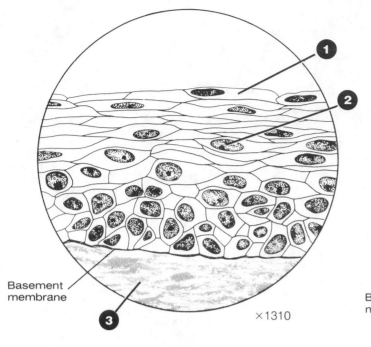

Basement membrane

×1310

■ **Figure 2.5** _____

_____

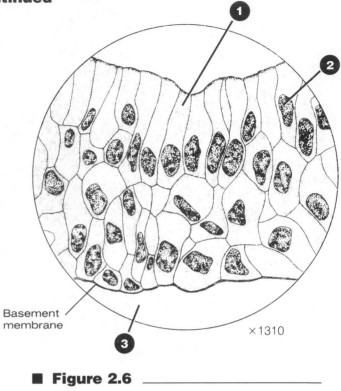

Basement membrane

×1310

■ **Figure 2.6** _____

_____

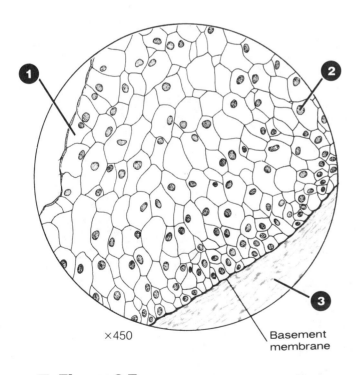

×450

Basement membrane

■ **Figure 2.7** _____

_____

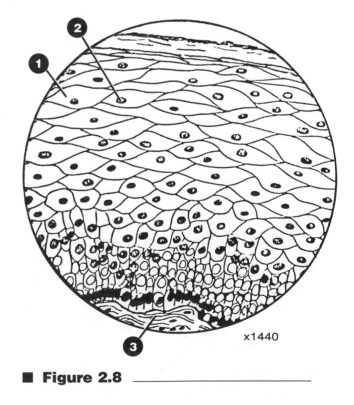

×1440

■ **Figure 2.8** _____

_____

■ **Figures 2.9–2.15: Review of epithelial tissues**

IDENTIFY

**Columnar cells
Cuboidal cells
Pseudostratified
    columnar cells
Squamous cells
Transitional cells**

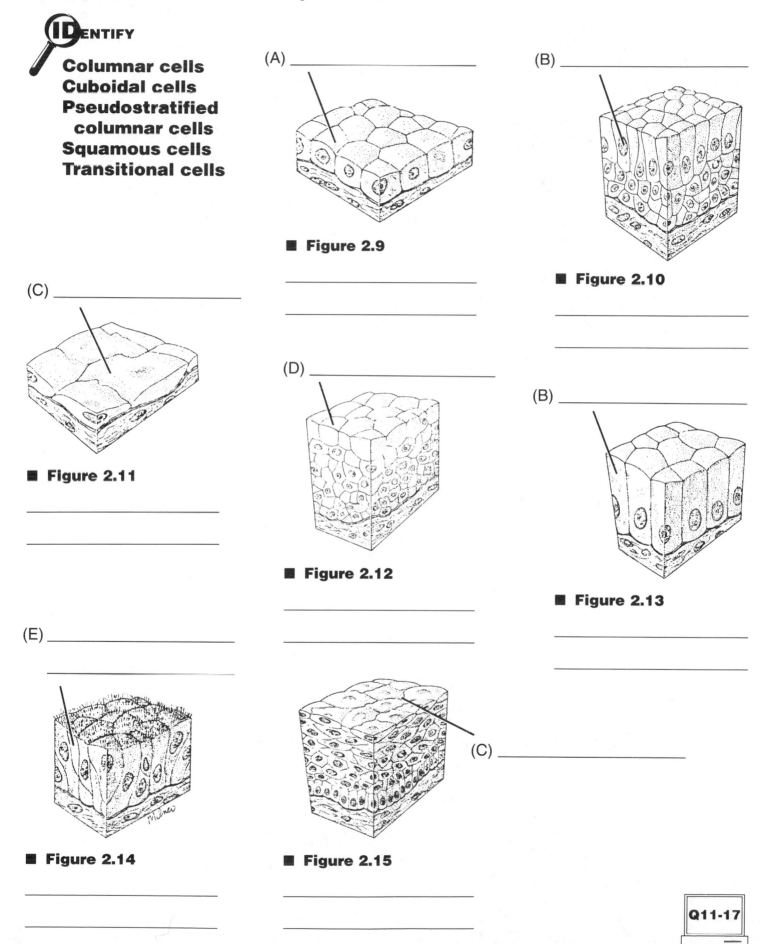

(A) _____

■ **Figure 2.9**

_____

_____

(B) _____

■ **Figure 2.10**

_____

_____

(C) _____

■ **Figure 2.11**

_____

_____

(D) _____

■ **Figure 2.12**

_____

_____

(B) _____

■ **Figure 2.13**

_____

_____

(E) _____

_____

■ **Figure 2.14**

_____

_____

■ **Figure 2.15**

(C) _____

_____

_____

■ Figures 2.16–2.19: Connective tissues

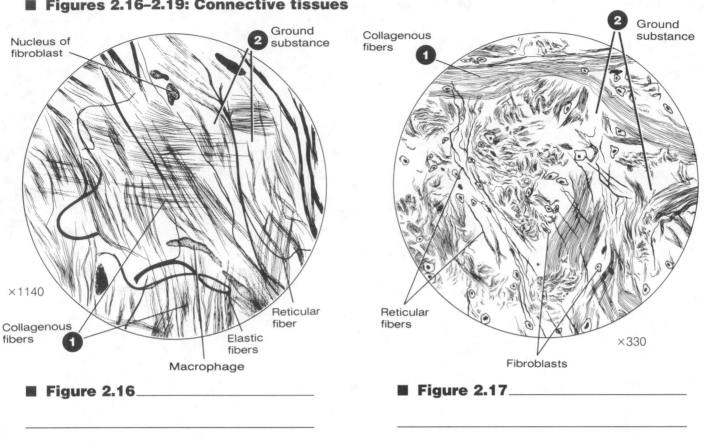

Nucleus of fibroblast

Ground substance — 2

×1140

Collagenous fibers — 1

Elastic fibers

Macrophage

Reticular fiber

Collagenous fibers

Ground substance — 2

1

Reticular fibers

Fibroblasts

×330

■ **Figure 2.16**_____

_____

■ **Figure 2.17**_____

_____

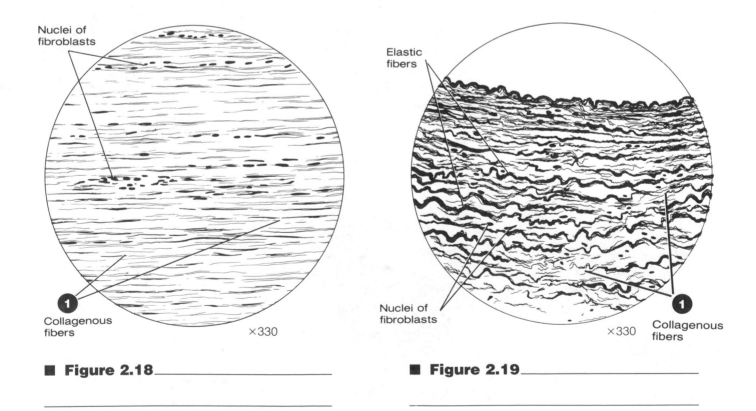

Nuclei of fibroblasts

1

Collagenous fibers

×330

Elastic fibers

Nuclei of fibroblasts

Collagenous fibers — 1

×330

■ **Figure 2.18**_____

_____

■ **Figure 2.19**_____

_____

Q18-22

■ Figures 2.20–2.24: Connective tissues, continued

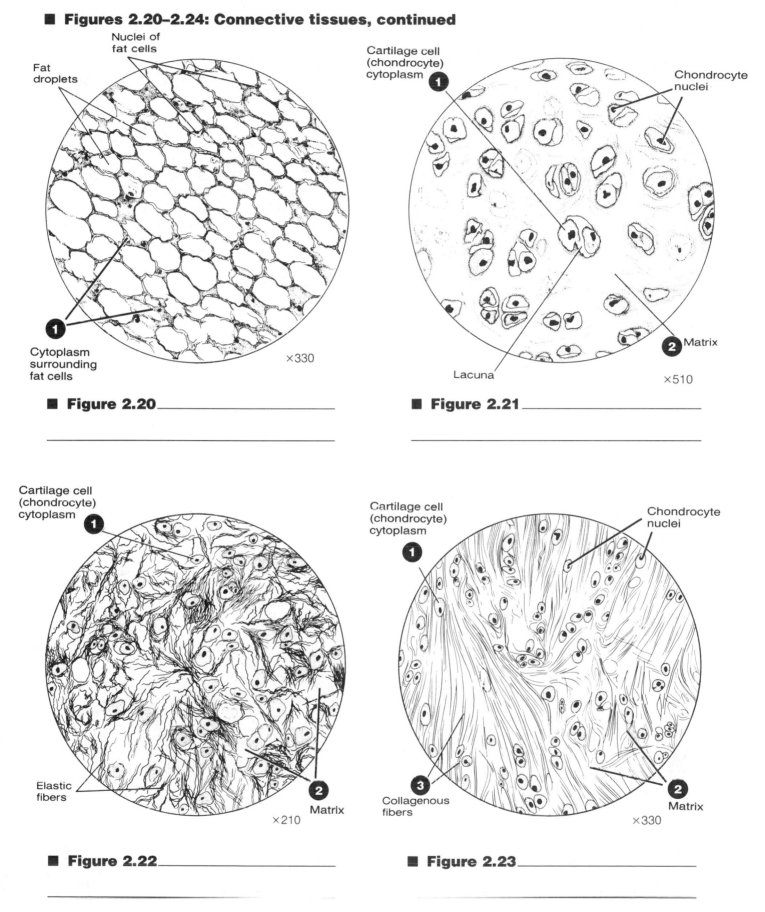

Nuclei of fat cells

Fat droplets

Cytoplasm surrounding fat cells

1

×330

■ Figure 2.20 _____

_____

Cartilage cell (chondrocyte) cytoplasm

1

Chondrocyte nuclei

Lacuna

2 Matrix

×510

■ Figure 2.21 _____

_____

Cartilage cell (chondrocyte) cytoplasm

1

Elastic fibers

2 Matrix

×210

■ Figure 2.22 _____

_____

Cartilage cell (chondrocyte) cytoplasm

1

Chondrocyte nuclei

3

Collagenous fibers

2 Matrix

×330

■ Figure 2.23 _____

_____

Q23-26

■ Figures 2.25–2.27: Review of connective tissues

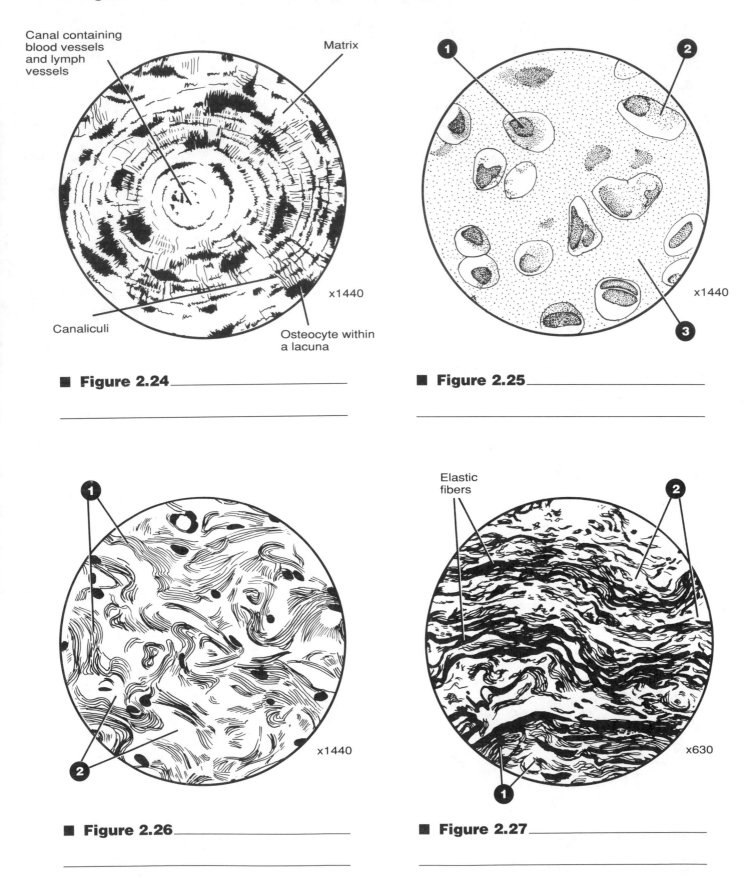

Canal containing
blood vessels
and lymph
vessels

Matrix

Canaliculi

Osteocyte within
a lacuna

x1440

■ **Figure 2.24**_____

_____

x1440

■ **Figure 2.25**_____

_____

x1440

■ **Figure 2.26**_____

_____

Elastic
fibers

x630

■ **Figure 2.27**_____

_____

■ Figures 2.28–2.31: Connective tissue review, continued

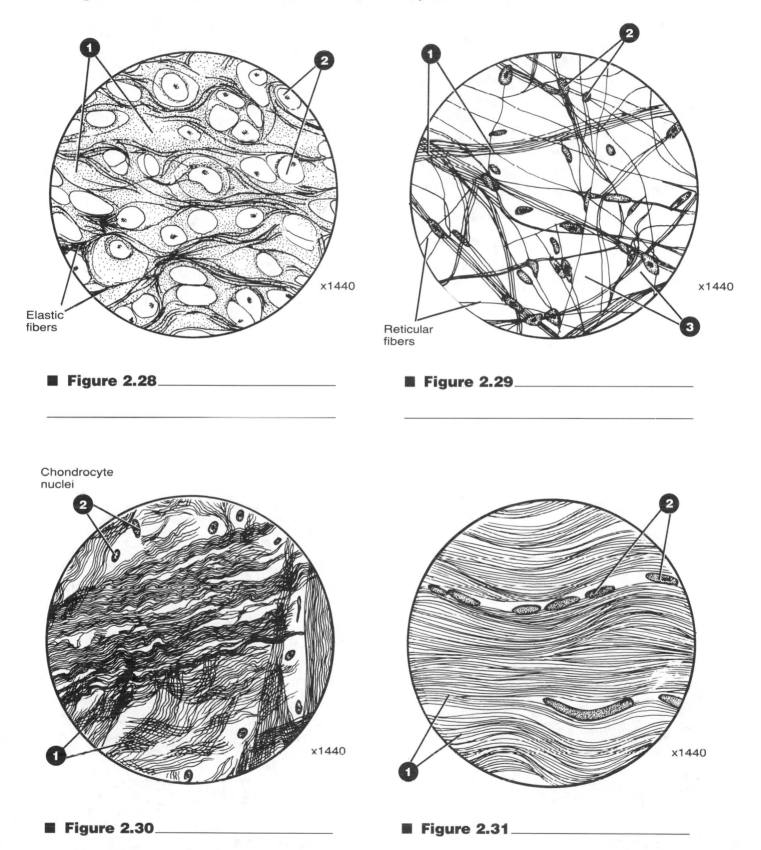

Elastic
fibers

×1440

■ Figure 2.28 _____

_____

Reticular
fibers

×1440

■ Figure 2.29 _____

_____

Chondrocyte
nuclei

×1440

■ Figure 2.30 _____

_____

×1440

■ Figure 2.31 _____

_____

Q31-34

■ Figures 2.32–2.35: Muscle tissues

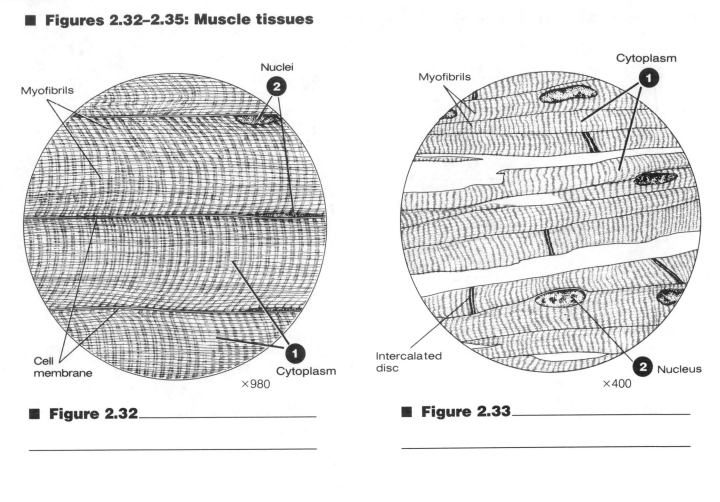

Myofibrils

Nuclei
2

Cell
membrane

1
Cytoplasm

×980

■ **Figure 2.32**_____

_____

Myofibrils

Cytoplasm
1

Intercalated
disc

2 Nucleus

×400

■ **Figure 2.33**_____

_____

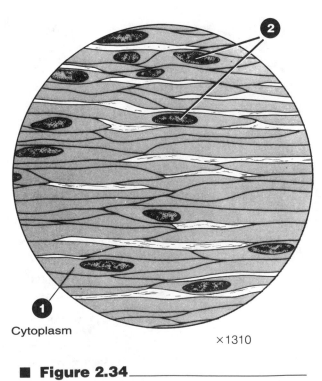

2

1
Cytoplasm

×1310

■ **Figure 2.34**_____

_____

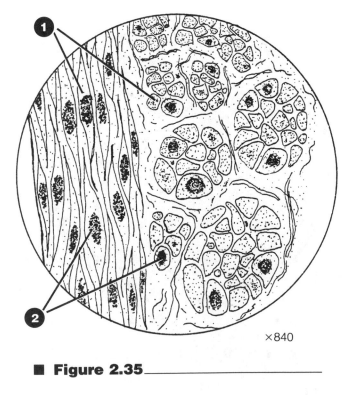

1

2

×840

■ **Figure 2.35**_____

_____

Q35-43

■ Figures 2.36–2.39: Muscle and nervous tissues

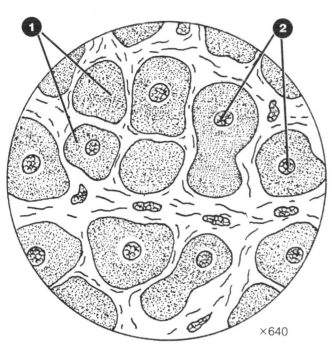

×640

■ Figure 2.36 _____

_____

Nuclei

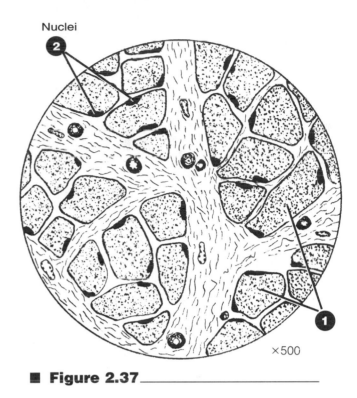

×500

■ Figure 2.37 _____

_____

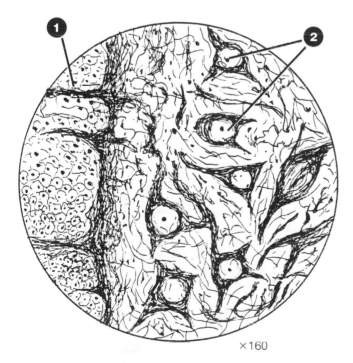

×160

■ Figure 2.38 _____

_____

■ Figure 2.39 _____

_____

Q41-45

MODULE 3

The Skeleton

Bones and Surface Markings

■ Table 3.1 Bone Markings

Projections Associated With Tendon or Ligament Attachment

Name	Description and/or Example
Process	
Crest	
Spine	
Epicondyle	
Tubercle	
Tuberosity	
Trochanter	

Projections That Help Form Joints

Condyle	
Facet	

Depressions and Openings

Fissure	
Foramen	
Meatus	
Sinus	
Sulcus	

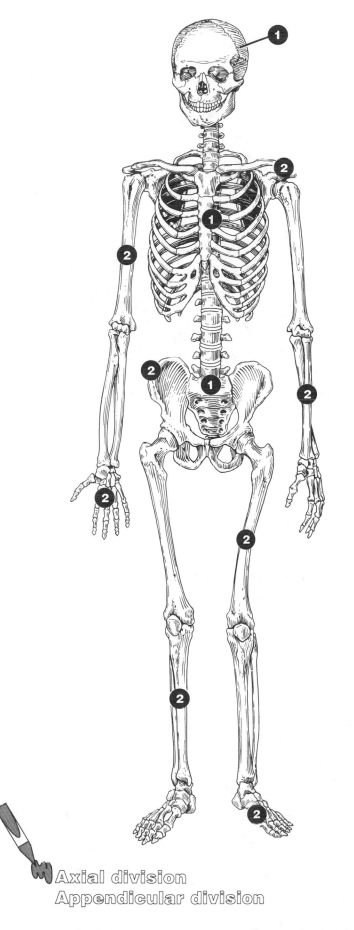

Axial division
Appendicular division

■ **Figure 3.1 Divisions of the skeleton**

Q11

IDENTIFY

Ethmoid
Frontal
Occipital
Parietal
Sphenoid
Temporal

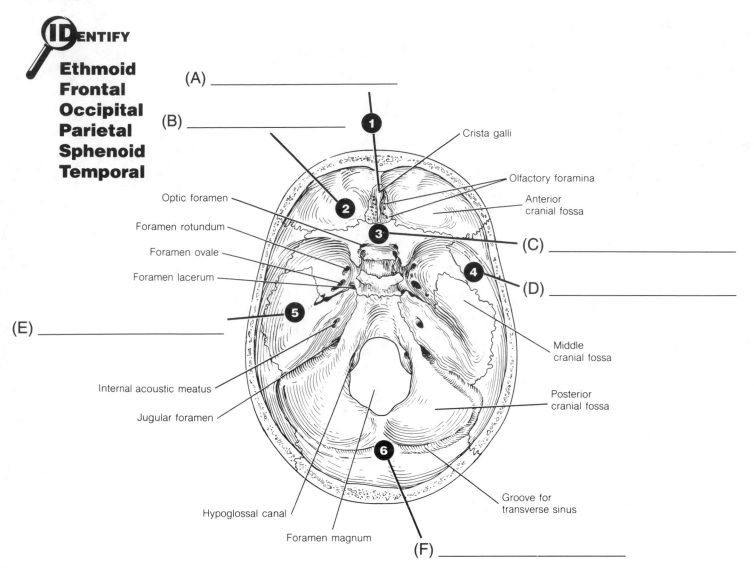

(A) _____

(B) _____

Crista galli

Olfactory foramina

Optic foramen

Anterior
cranial fossa

Foramen rotundum

(C) _____

Foramen ovale

Foramen lacerum

(D) _____

(E) _____

Middle
cranial fossa

Internal acoustic meatus

Posterior
cranial fossa

Jugular foramen

Groove for
transverse sinus

Hypoglossal canal

Foramen magnum

(F) _____

■ Figure 3.2 Transverse section through cranial cavity

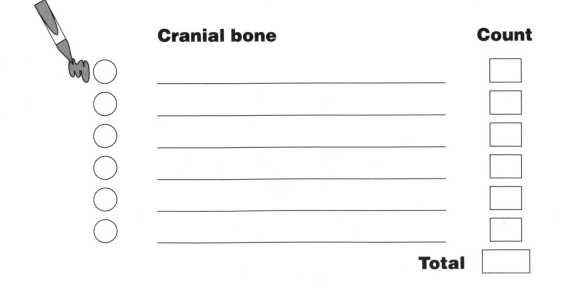

Cranial bone	Count
_____	☐
_____	☐
_____	☐
_____	☐
_____	☐
_____	☐
Total	☐

IDENTIFY

Inferior nasal conchae
Lacrimal
Mandible
Maxillae
Nasal
Vomer
Zygomatic

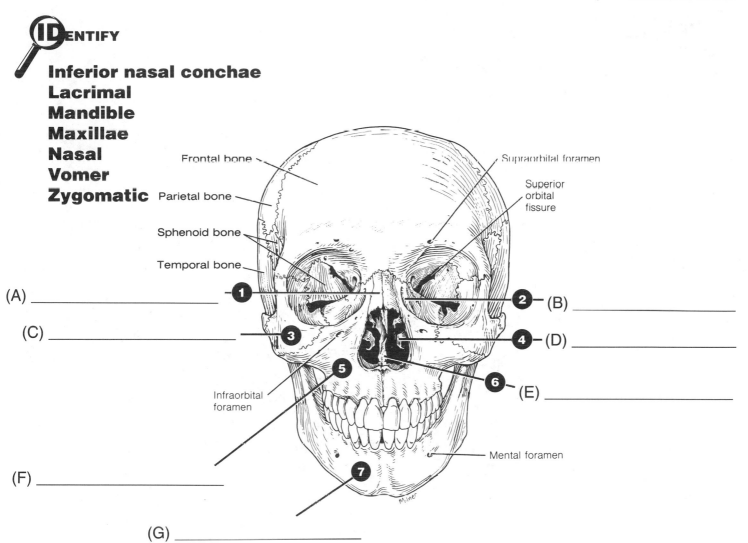

Frontal bone

Parietal bone

Sphenoid bone

Temporal bone

Supraorbital foramen

Superior orbital fissure

(A) _____

1

(C) _____

3

5

Infraorbital foramen

(F) _____

2 —(B) _____

4 —(D) _____

6 —(E) _____

Mental foramen

7

(G) _____

■ **Figure 3.3 Anterior view of the skull**

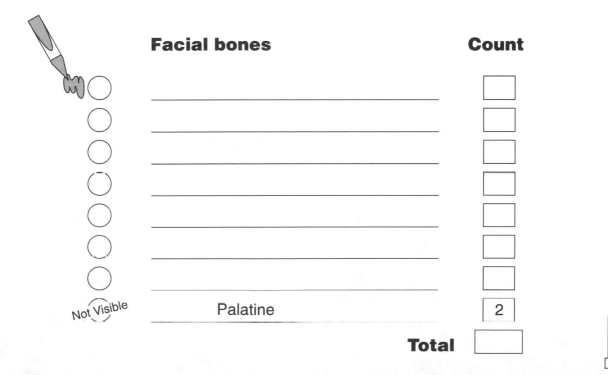

Facial bones

Count

○ _____ ☐

○ _____ ☐

○ _____ ☐

○ _____ ☐

○ _____ ☐

○ _____ ☐

○ _____ ☐

Not Visible   Palatine   2

Total ☐

Q13-50

IDENTIFY

Inferior orbital fissure
Superior orbital fissure

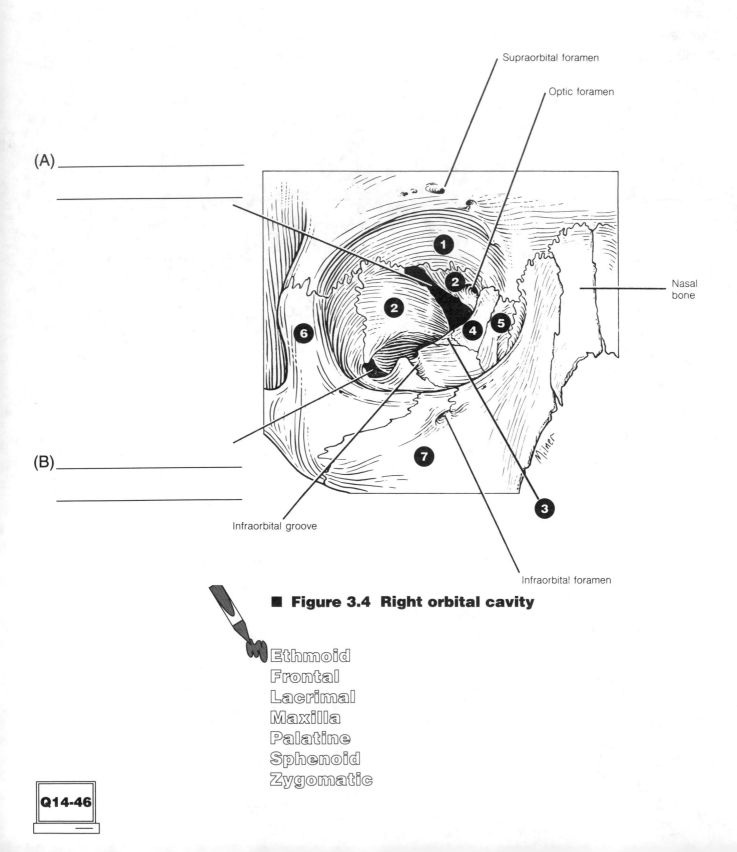

(A) _____

_____

(B) _____

_____

■ **Figure 3.4 Right orbital cavity**

Ethmoid
Frontal
Lacrimal
Maxilla
Palatine
Sphenoid
Zygomatic

Q14-46

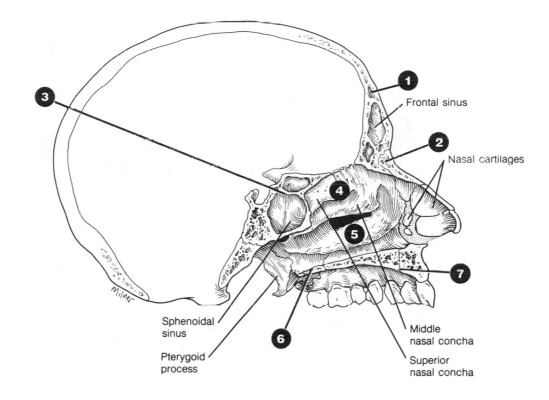

■ Figure 3.5 Left lateral (internal) view of nasal cavity

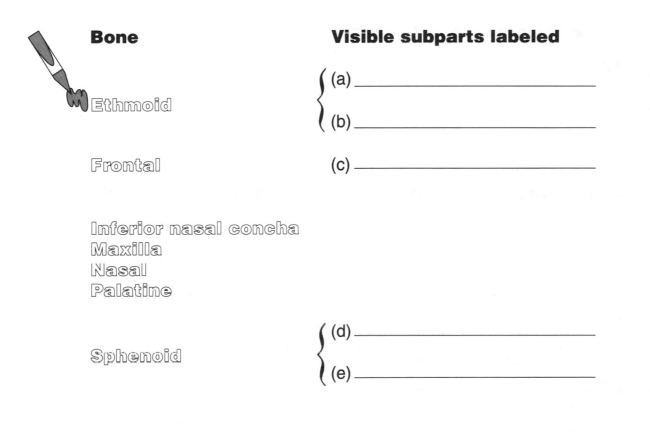

Bone	Visible subparts labeled
Ethmoid	(a) _____ (b) _____
Frontal	(c) _____
Inferior nasal concha Maxilla Nasal Palatine	
Sphenoid	(d) _____ (e) _____

Q15-17

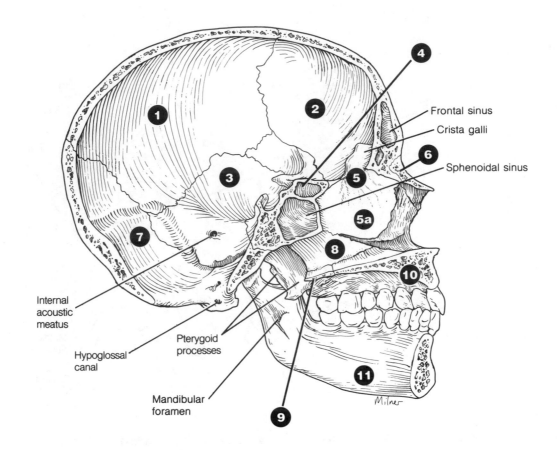

Frontal sinus
Crista galli
Sphenoidal sinus

Internal
acoustic
meatus

Hypoglossal
canal

Pterygoid
processes

Mandibular
foramen

Milner

■ Figure 3.6 Sagittal view of the skull

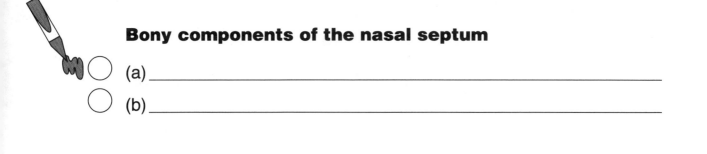

Bony components of the nasal septum

○ (a) _____

○ (b) _____

Cranial bones (6 visible)

○ (c) _____

○ (d) _____

○ (e) _____

○ (f) _____

○ (g) _____

○ (h) _____

Facial bones (5 visible)

○ (i) _____

○ (j) _____

○ (k) _____

○ (l) _____

○ (m) _____

Q18-47

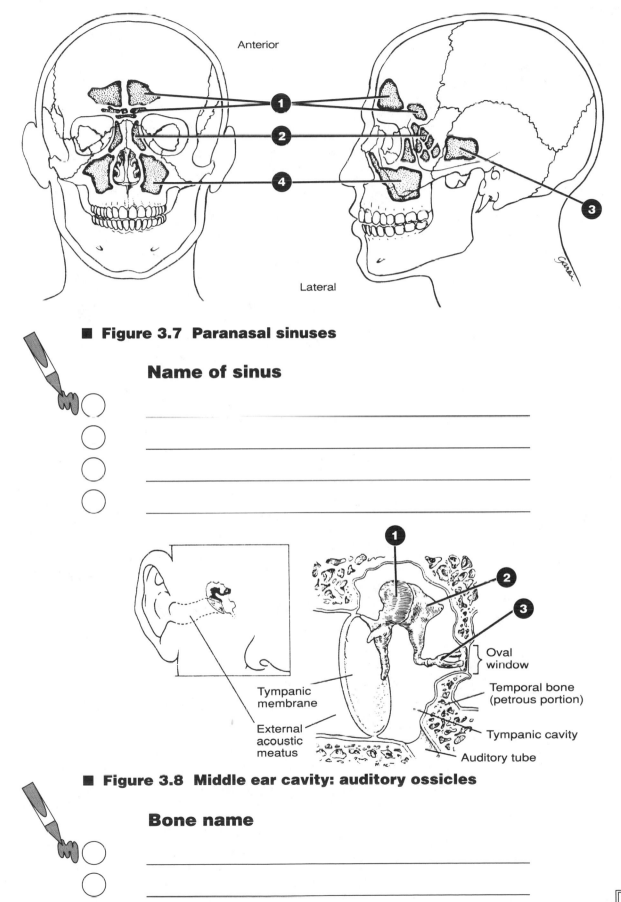

Anterior

Lateral

■ Figure 3.7 Paranasal sinuses

Name of sinus

○ _____

○ _____

○ _____

○ _____

Tympanic
membrane

External
acoustic
meatus

Oval
window

Temporal bone
(petrous portion)

Tympanic cavity

Auditory tube

■ Figure 3.8 Middle ear cavity: auditory ossicles

Bone name

○ _____

○ _____

○ _____

Q19-20

IDENTIFY

Coronal suture
Lambdoidal suture
Squamosal suture

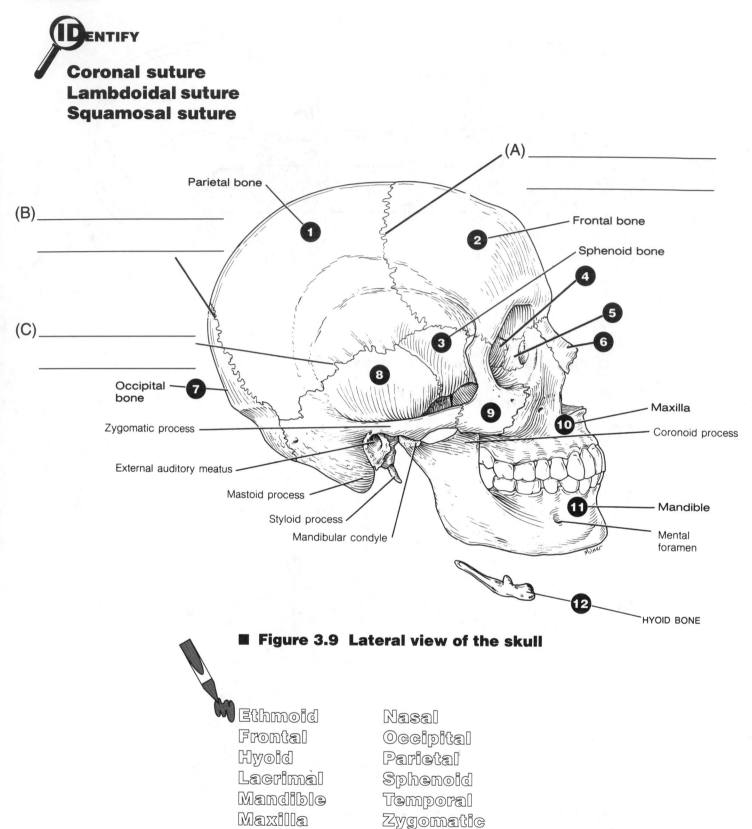

(A) _____

_____

(B) _____

_____

(C) _____

_____

Parietal bone

Frontal bone

Sphenoid bone

Occipital bone

Zygomatic process

External auditory meatus

Mastoid process

Styloid process

Mandibular condyle

Maxilla

Coronoid process

Mandible

Mental foramen

HYOID BONE

1 2 3 4 5 6 7 8 9 10 11 12

■ **Figure 3.9 Lateral view of the skull**

Ethmoid   Nasal
Frontal   Occipital
Hyoid   Parietal
Lacrimal   Sphenoid
Mandible   Temporal
Maxilla   Zygomatic

IDENTIFY

Incisive foramen
Jugular foramen
Mastoid process
Stylomastoid foramen
Occipital condyle

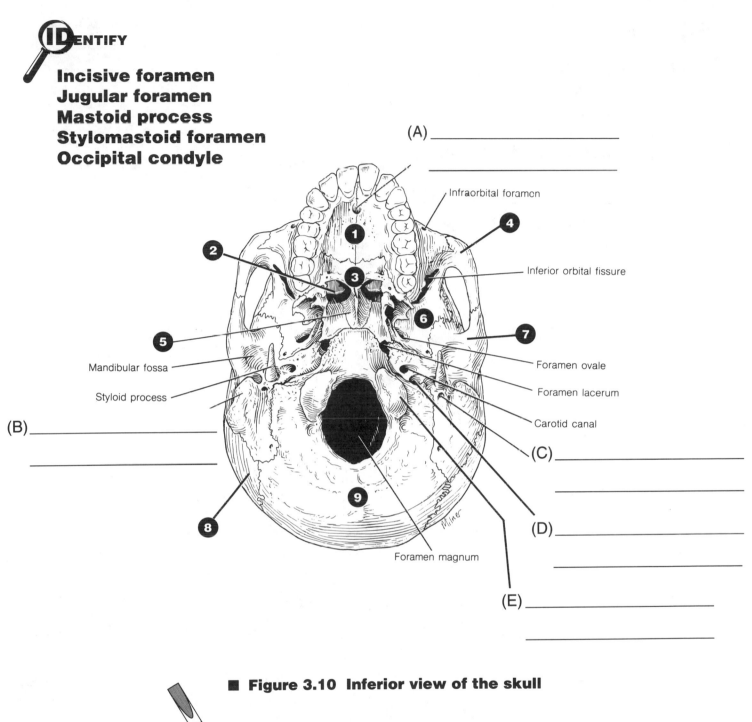

(A) _____

_____

Infraorbital foramen

Inferior orbital fissure

Mandibular fossa

Styloid process

(B) _____

_____

Foramen ovale

Foramen lacerum

Carotid canal

(C) _____

_____

(D) _____

_____

Foramen magnum

(E) _____

_____

■ **Figure 3.10 Inferior view of the skull**

Inferior nasal conchae
Maxilla
Occipital
Palatine
Parietal
Sphenoid
Temporal
Vomer
Zygomatic

Q25-49

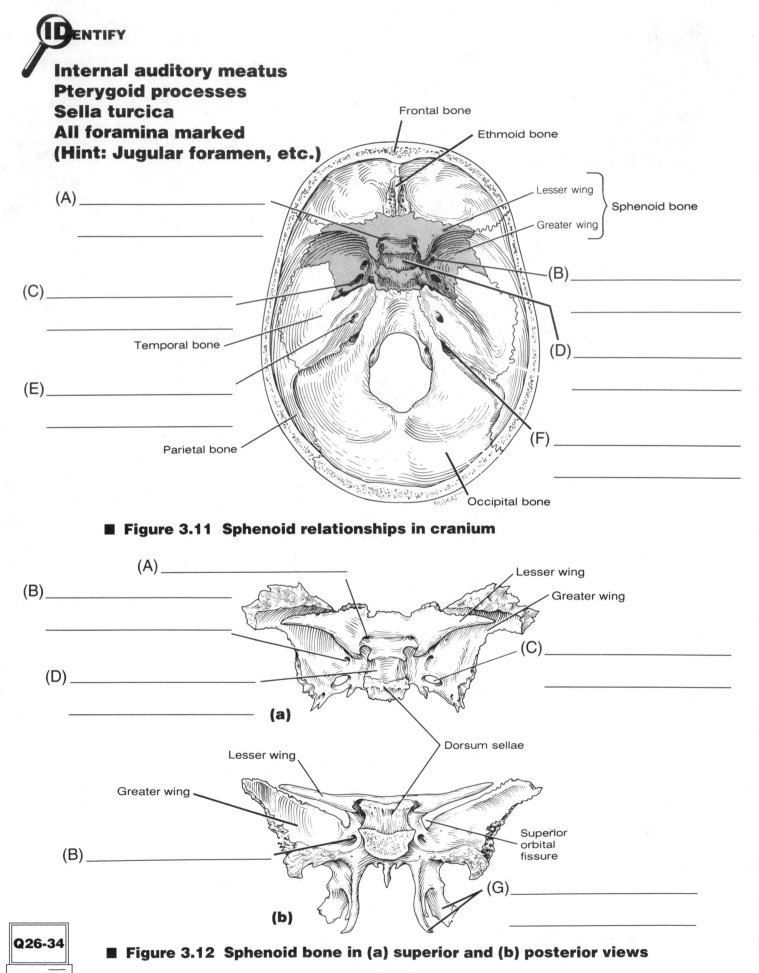

IDENTIFY

**Internal auditory meatus
Pterygoid processes
Sella turcica
All foramina marked
(Hint: Jugular foramen, etc.)**

(A)_____

_____

(C)_____

_____

(E)_____

_____

Frontal bone

Ethmoid bone

Lesser wing

Greater wing

Sphenoid bone

Temporal bone

Parietal bone

Occipital bone

(B)_____

_____

(D)_____

_____

(F)_____

_____

■ **Figure 3.11 Sphenoid relationships in cranium**

(A)_____

(B)_____

_____

(D)_____

_____

(a)

Lesser wing

Greater wing

(C)_____

_____

Dorsum sellae

Lesser wing

Greater wing

(B)_____

Superior
orbital
fissure

(G)_____

_____

(b)

■ **Figure 3.12 Sphenoid bone in (a) superior and (b) posterior views**

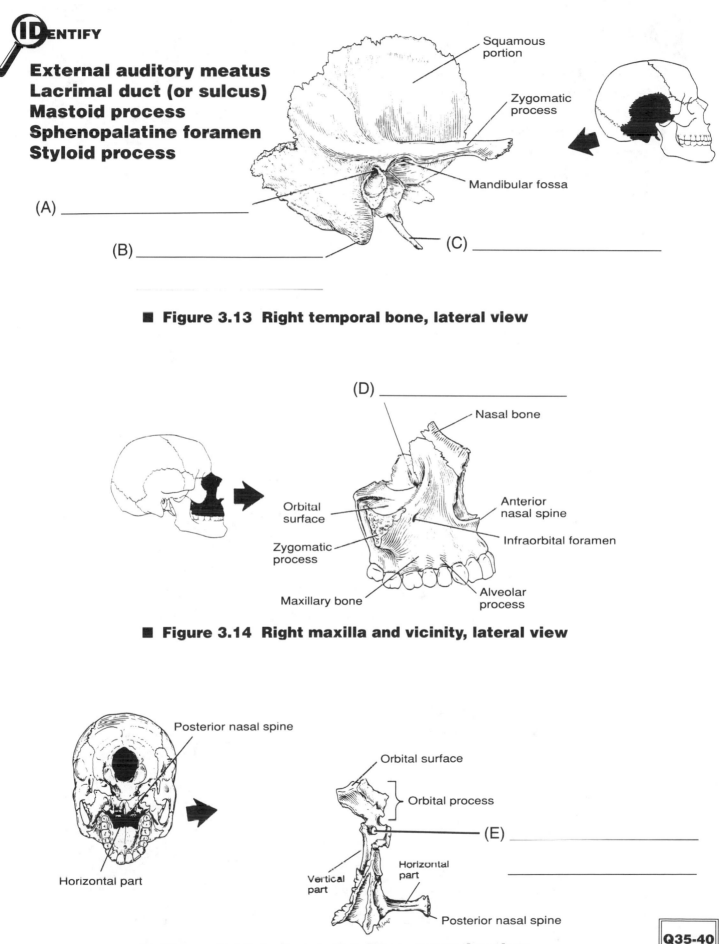

IDENTIFY

External auditory meatus
Lacrimal duct (or sulcus)
Mastoid process
Sphenopalatine foramen
Styloid process

Squamous portion

Zygomatic process

Mandibular fossa

(A) _____

(B) _____

(C) _____

■ **Figure 3.13 Right temporal bone, lateral view**

(D) _____

Nasal bone

Orbital surface

Zygomatic process

Maxillary bone

Anterior nasal spine

Infraorbital foramen

Alveolar process

■ **Figure 3.14 Right maxilla and vicinity, lateral view**

Posterior nasal spine

Orbital surface

Orbital process

(E) _____

_____

Horizontal part

Vertical part

Horizontal part

Posterior nasal spine

■ **Figure 3.15 Left palatine bone, posterior view**

Q35-40

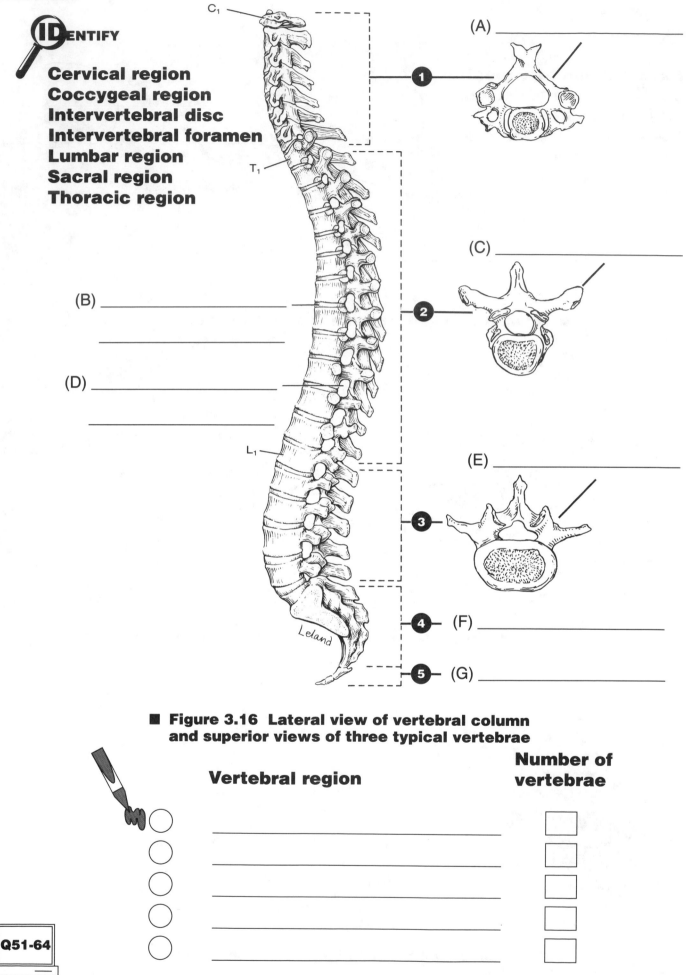

IDENTIFY

Cervical region
Coccygeal region
Intervertebral disc
Intervertebral foramen
Lumbar region
Sacral region
Thoracic region

C_1

T_1

L_1

Leland

(A) _____

(C) _____

(E) _____

(B) _____
_____
(D) _____
_____

(F) _____

(G) _____

1

2

3

4

5

■ **Figure 3.16 Lateral view of vertebral column
and superior views of three typical vertebrae**

Vertebral region

**Number of
vertebrae**

Q51-64

IDENTIFY

**All structures marked
(Hint: Lamina, Transverse process,
Superior articular process, etc.)**

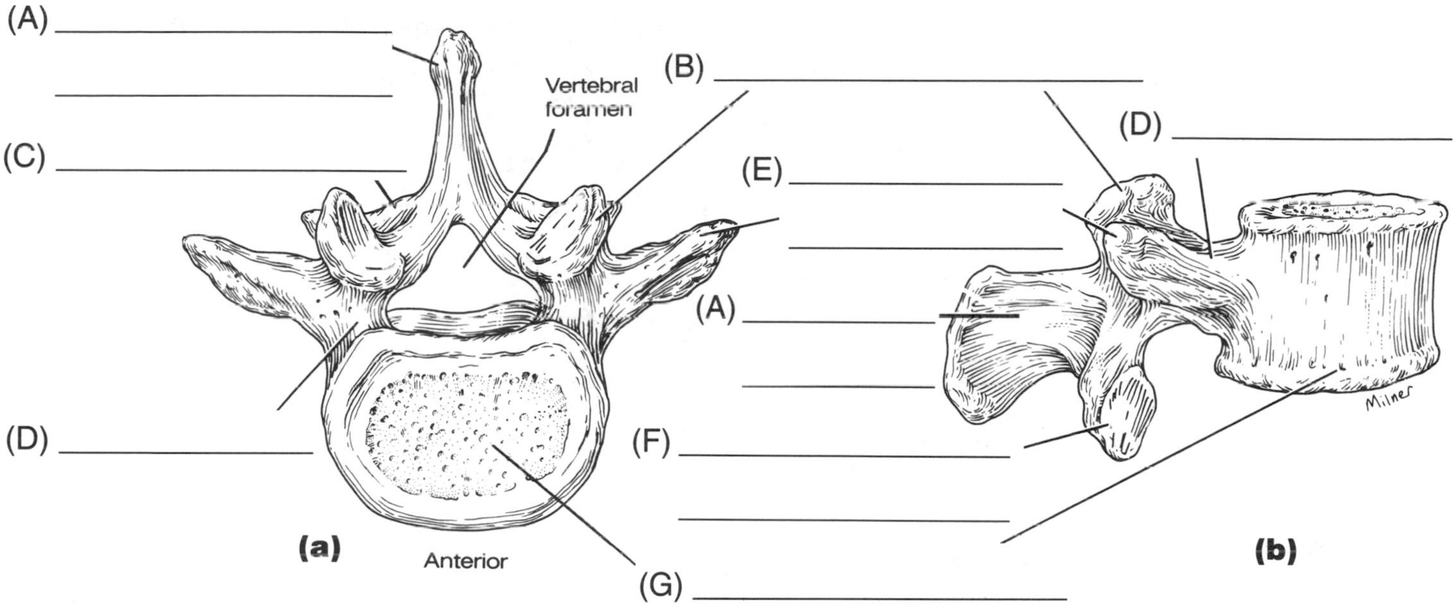

(A) _____

_____

(C) _____

Vertebral foramen

(B) _____

(D) _____

(E) _____

(A) _____

(F) _____

(D) _____

(a)

Anterior

(G) _____

(b)

Milner

■ **Figure 3.17 Typical vertebra in (a) superior and (b) lateral views**

Atlas   Lamina
Axis   Pedicle
Centrum (body)   Spinous process

1
2
Superior articular process
3
Vertebral foramen
Neural arch
4
Transverse foramen

Anterior arch
Superior articular surface for occipital condyles
Odontoid process (dens)
5
6
Transverse foramina
Transverse processes
Posterior arch
Spinous process
Milner

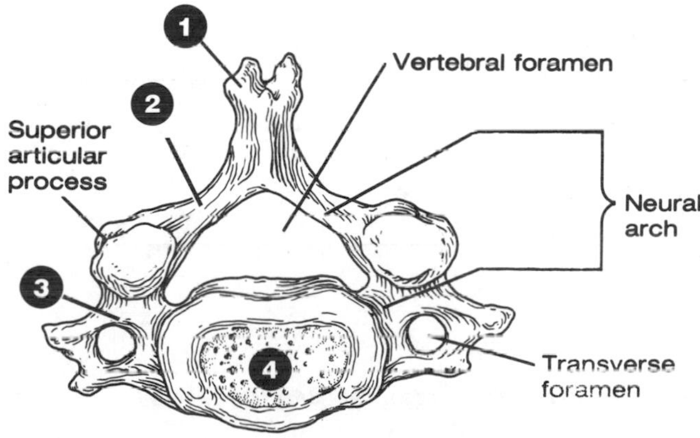

■ **Figure 3.18 Typical cervical vertebra, superior view**

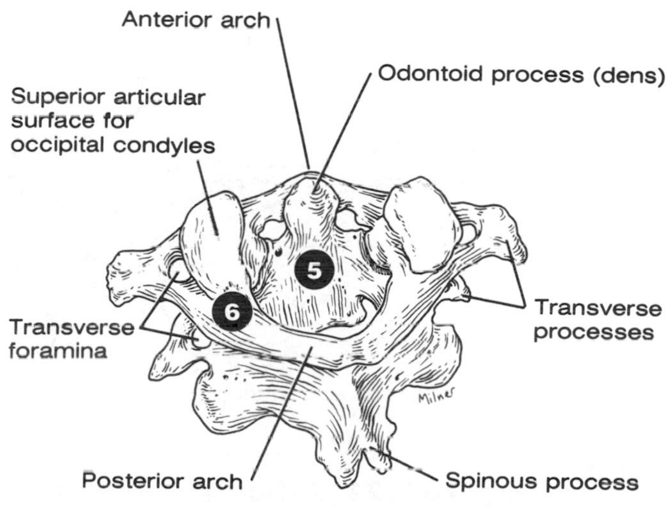

■ **Figure 3.19 First and second cervical vertebrae**

Q55-63

IDENTIFY

**(Articular) facet
Demifacet**

Body of vertebrae
Inferior articular
     processes
Superior articular
     processes

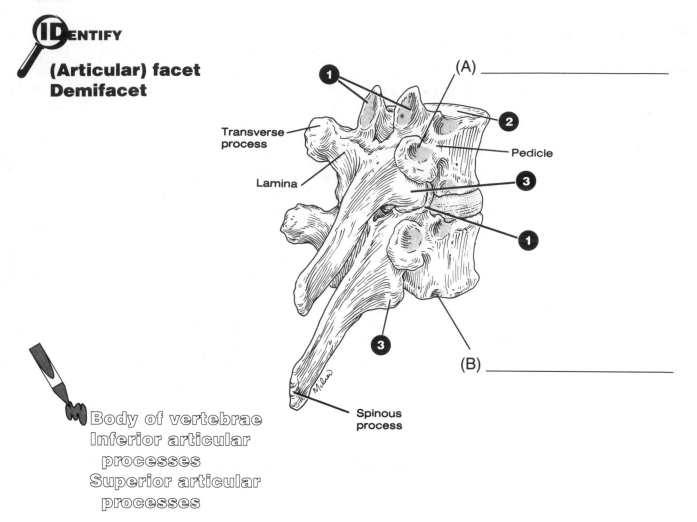

(A) _____

Transverse
process

Lamina

Pedicle

(B) _____

Spinous
process

■ **Figure 3.20 Two typical thoracic vertebrae, posterolateral view**

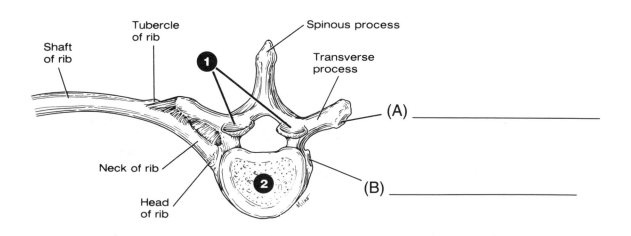

Shaft
of rib

Tubercle
of rib

Spinous process

Transverse
process

(A) _____

Neck of rib

Head
of rib

(B) _____

■ **Figure 3.21 Articulation of rib and thoracic vertebra, superior view**

Q65-66

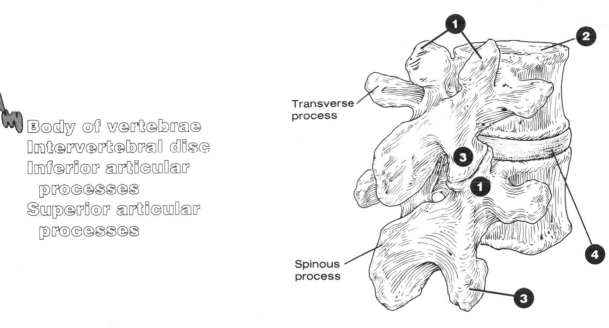

Body of vertebrae
Intervertebral disc
Inferior articular
    processes
Superior articular
    processes

Transverse process

Spinous process

■ Figure 3.22 Two typical lumbar vertebrae, posterolateral view

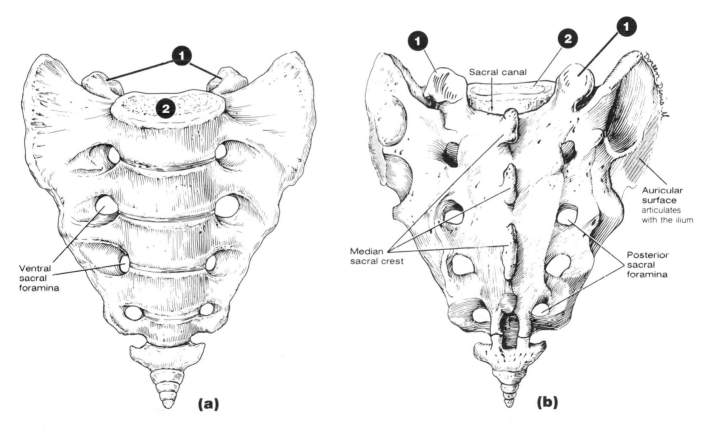

Sacral canal

Auricular surface articulates with the ilium

Ventral sacral foramina

Median sacral crest

Posterior sacral foramina

(a)

(b)

■ Figure 3.23 Sacrum and coccyx in (a) anterior and (b) posterior views

Q54-67

IDENTIFY

**Parts of the sternum
(Hint: Body, Manubrium, etc.)
Parts of a rib
(Hint: Head, Neck, etc.)**

(A) _____

(B) _____

(C) _____

Costal cartilage
False ribs
True ribs

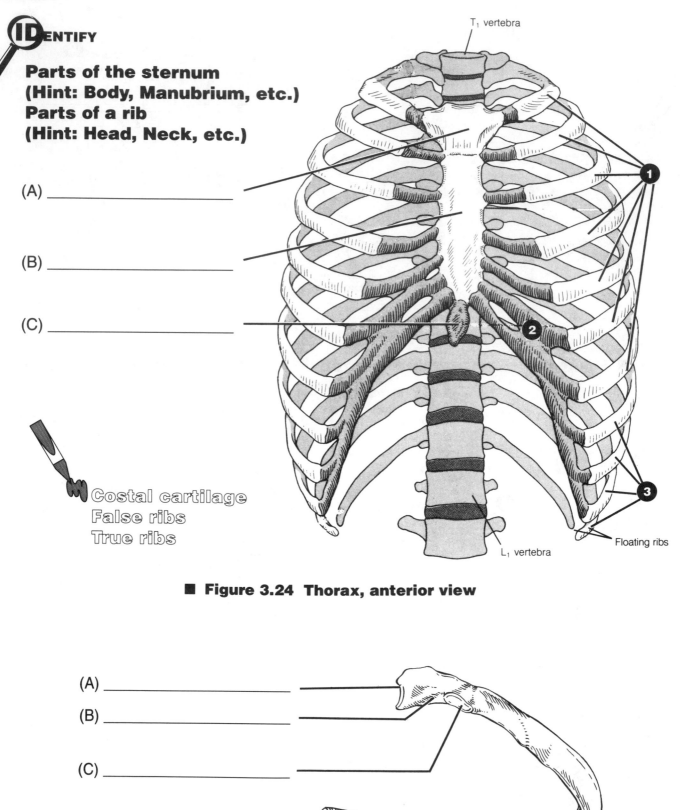

T₁ vertebra

1

2

3

Floating ribs

L₁ vertebra

■ **Figure 3.24 Thorax, anterior view**

(A) _____ _____

(B) _____

(C) _____

(D) _____

■ **Figure 3.25 Right rib, posterior view**

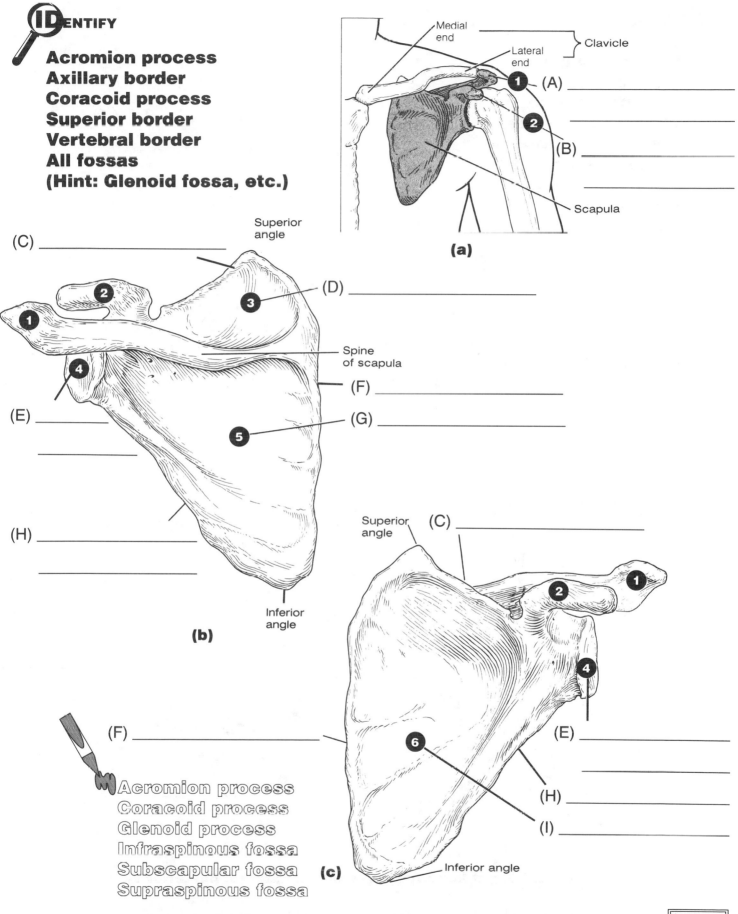

IDENTIFY

Acromion process
Axillary border
Coracoid process
Superior border
Vertebral border
All fossas
(Hint: Glenoid fossa, etc.)

(a) Left shoulder girdle, anterior view

Medial end
Lateral end
Clavicle
1 (A) _____
2 _____
(B) _____
_____
Scapula

(a)

Superior angle
(C) _____
2
1
3
(D) _____
Spine of scapula
4
(F) _____
(E) _____
5
(G) _____
_____
(H) _____
_____
Inferior angle
(b)

Superior angle
(C) _____
2
1
4
6
(E) _____
_____
(H) _____
(I) _____
(F) _____

Acromion process
Coracoid process
Glenoid process
Infraspinous fossa
Subscapular fossa
Supraspinous fossa
(c)
Inferior angle

■ **Figure 3.26 (a) Left shoulder girdle, anterior view, (b) left**
scapula, posterior view, and (c) left scapula, anterior view

Q71-77

IDENTIFY

Anatomical neck
Capitulum
Greater tubercle
Lateral epicondyle
Lesser tubercle
Medial epicondyle
Olecranon fossa
Surgical neck
Trochlea

Capitulum
Greater tubercle
Lateral epicondyle
Lesser tubercle
Medial epicondyle
Trochlea

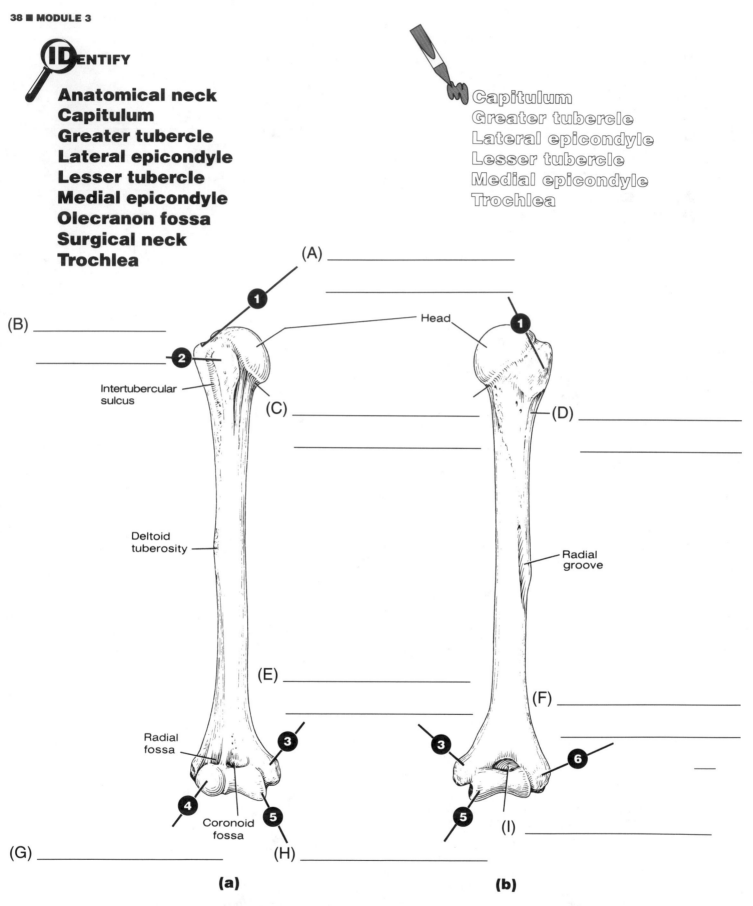

(A) _____

_____

(B) _____

_____ ②

Intertubercular
sulcus

(C) _____

_____

Head

① (b)

(D) _____

_____

Deltoid
tuberosity

Radial
groove

(E) _____

_____

(F) _____

_____

Radial
fossa

③

④

Coronoid
fossa

⑤

(G) _____

(H) _____

③

⑤

⑥

_____

(I) _____

(a)   **(b)**

■ **Figure 3.27 Right humerus in (a) anterior and (b) posterior views**

IDENTIFY

Coronoid process
Olecranon process
Styloid process of radius
Styloid process of ulna
Trochlear notch

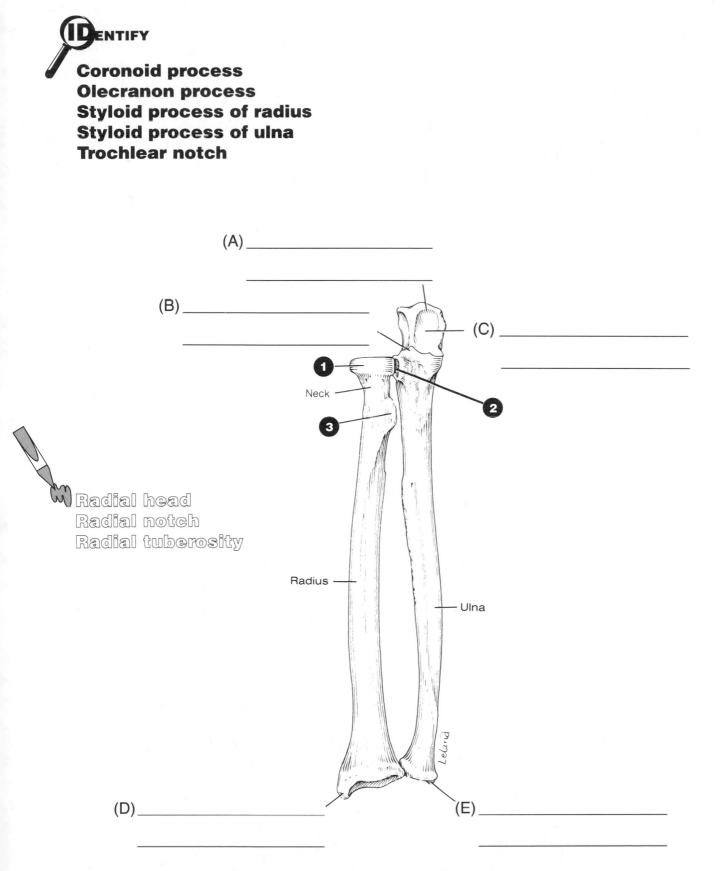

(A) _____

_____

(B) _____

_____

(C) _____

_____

1

Neck

3

2

Radial head
Radial notch
Radial tuberosity

Radius

Ulna

Leland

(D) _____

_____

(E) _____

_____

■ **Figure 3.28 Right radius and ulna, anterior view**

Q83-86

IDENTIFY

Carpals
Metacarpals
Phalanges:
  proximal
  middle
  distal
  4th middle phalanx

Capitate
Hamate
Lunate
Pisiform
Scaphoid
Trapezium
Trapezoid
Triquetral

(A) _____

(B) _____

(C) _____

(D) _____

(E) _____

(F) _____

(G) _____

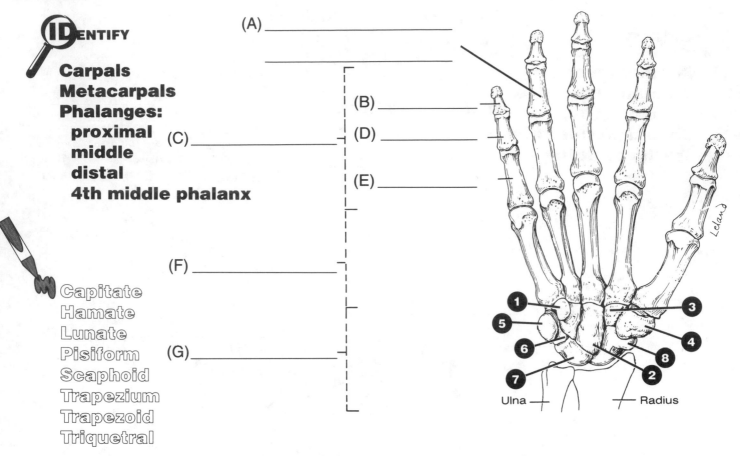

■ **Figure 3.29 Bones of the right wrist and hand, anterior view**

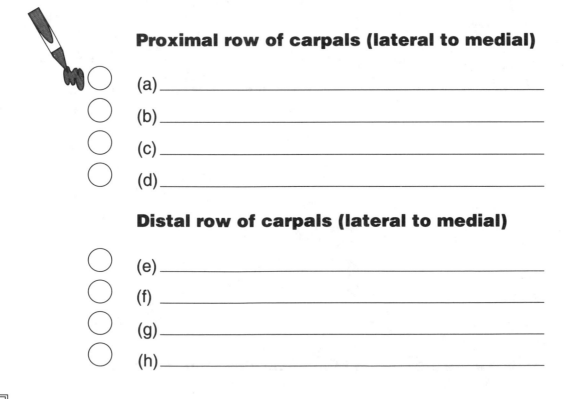

Proximal row of carpals (lateral to medial)

(a) _____

(b) _____

(c) _____

(d) _____

Distal row of carpals (lateral to medial)

(e) _____

(f) _____

(g) _____

(h) _____

Q87-89

IDENTIFY

Ischial spine
Sacroiliac symphysis

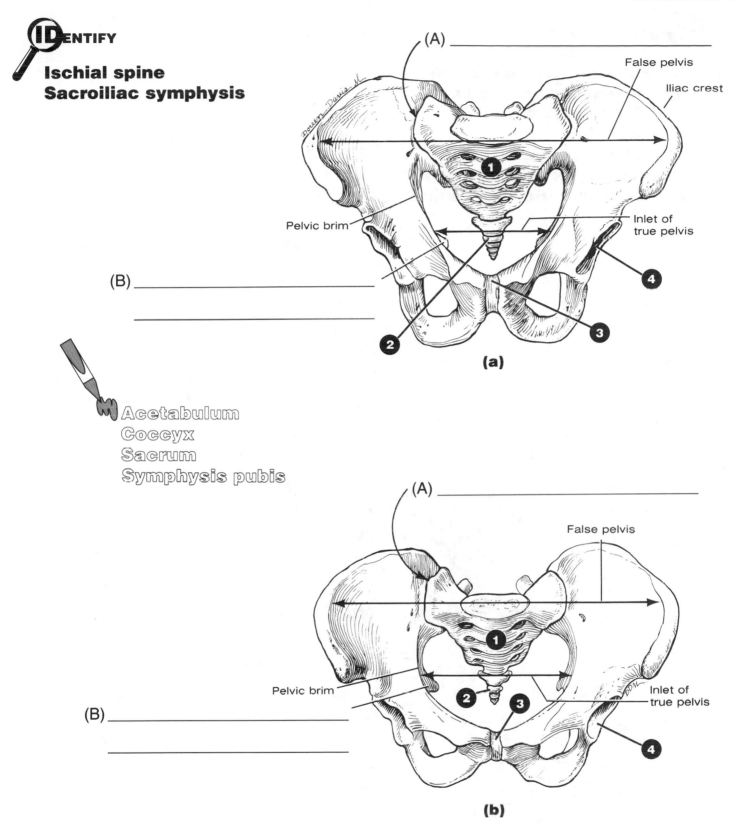

Acetabulum
Coccyx
Sacrum
Symphysis pubis

(a) male pelvis labels: (A), False pelvis, Iliac crest, Pelvic brim, Inlet of true pelvis, (B), 1, 2, 3, 4

(b) female pelvis labels: (A), False pelvis, Pelvic brim, Inlet of true pelvis, (B), 1, 2, 3, 4

■ **Figure 3.30 Anterior views of (a) male pelvis and (b) female pelvis**

Q90-96

IDENTIFY

Anterior inferior iliac spine
Anterior superior iliac spine
Greater sciatic notch
Ischial spine
Obturator foramen
Posterior inferior iliac spine
Posterior superior iliac spine

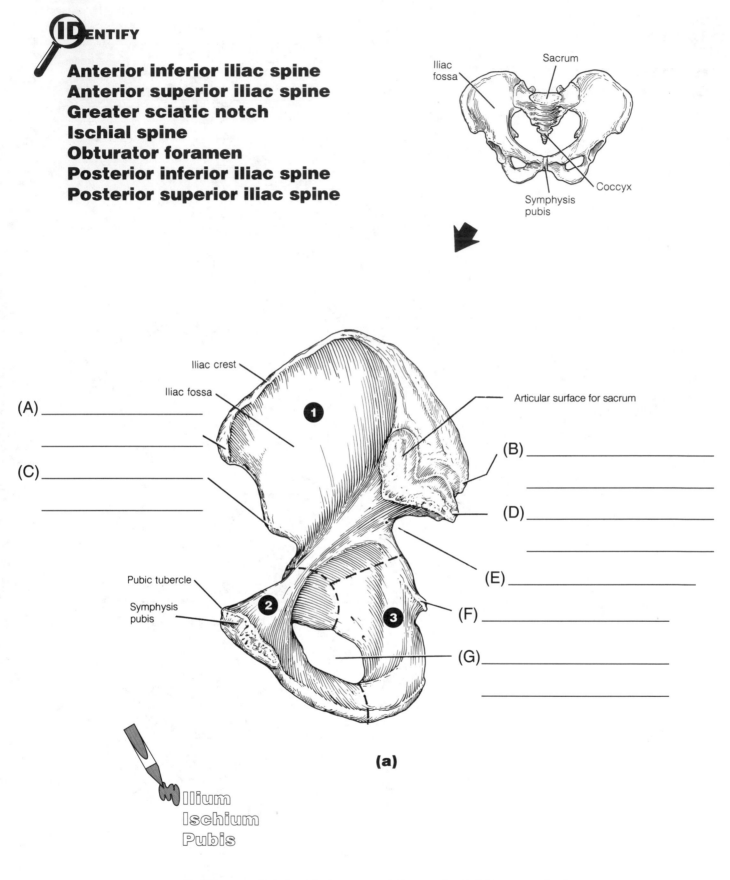

(A) _____

_____

(C) _____

_____

(B) _____

_____

(D) _____

_____

(E) _____

(F) _____

(G) _____

_____

(a)

Ilium
Ischium
Pubis

■ **Figure 3.31 Right coxal bone in (a) internal view**

Q90-94

IDENTIFY

Anterior inferior iliac spine
Anterior superior iliac spine
Greater sciatic notch
Ischial spine
Ischial tuberosity
Obturator foramen
Posterior inferior iliac spine
Posterior superior iliac spine

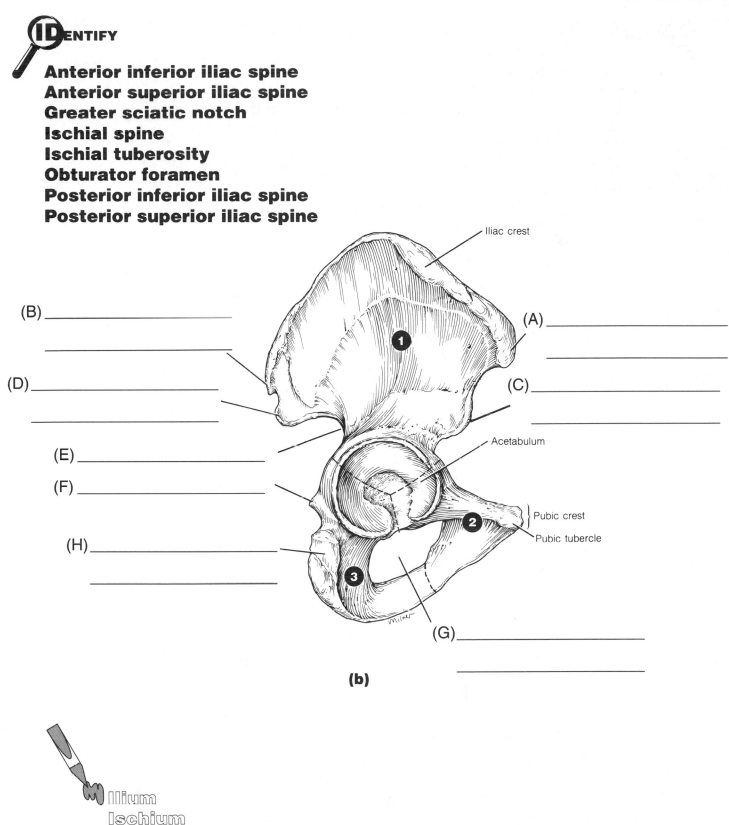

(B) _____

_____

(D) _____

_____

(E) _____

(F) _____

(H) _____

_____

Iliac crest

(A) _____

_____

(C) _____

_____

Acetabulum

Pubic crest

Pubic tubercle

(G) _____

_____

(b)

Illium
Ischium
Pubis

■ **Figure 3.31 continued: Right coxal bone in (b) external view**

IDENTIFY

Adductor tubercle
Fovea capitis
All epicondyles (Hint: Lateral epicondyle, etc.)

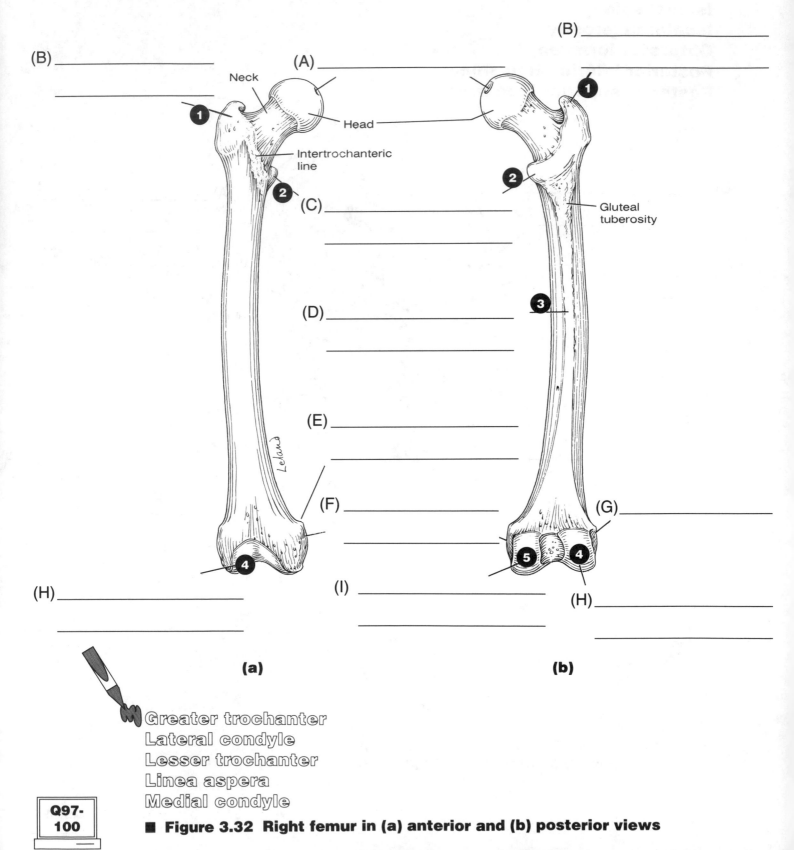

(B) _____

(A) _____
Neck

(B) _____
_____

Head

1

Intertrochanteric
line

1

2

(C) _____

_____

Gluteal
tuberosity

2

3

(D) _____

_____

(E) _____

_____

(F) _____

(G) _____

_____

5   4

4

(H) _____

(I) _____

(H) _____

_____

_____

_____

(a)

(b)

Greater trochanter
Lateral condyle
Lesser trochanter
Linea aspera
Medial condyle

Q97-
100

■ **Figure 3.32 Right femur in (a) anterior and (b) posterior views**

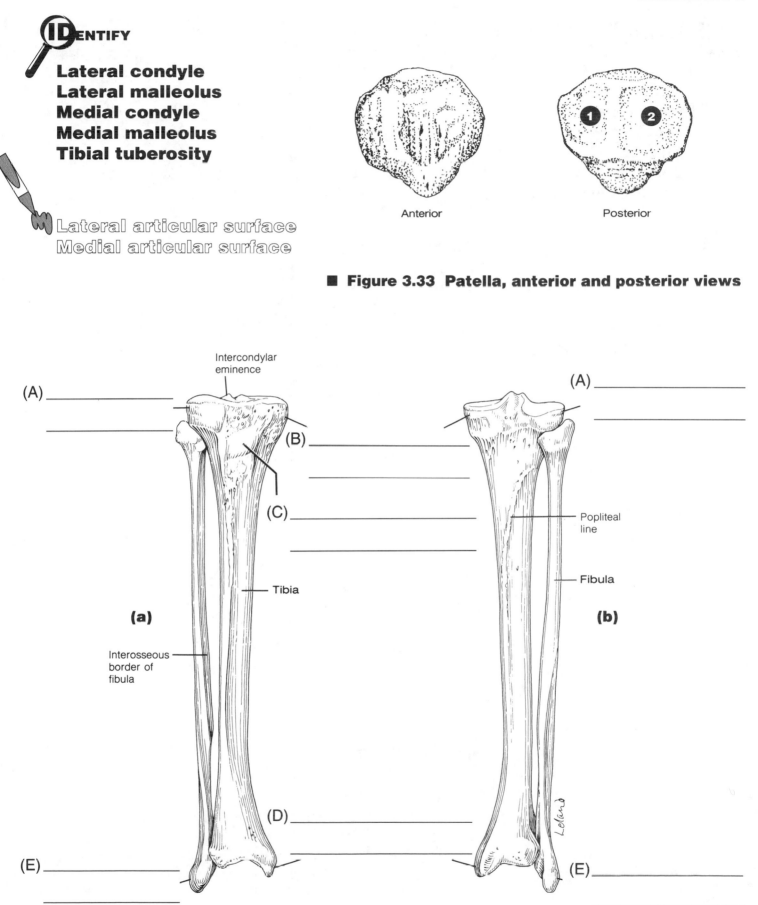

(ID)ENTIFY

Lateral condyle
Lateral malleolus
Medial condyle
Medial malleolus
Tibial tuberosity

Lateral articular surface
Medial articular surface

Anterior   Posterior

■ **Figure 3.33 Patella, anterior and posterior views**

Intercondylar
eminence

(A) _____

_____

(B) _____

_____

(C) _____

_____

(A) _____

_____

Popliteal
line

Fibula

Tibia

(a)

(b)

Interosseous
border of
fibula

(D) _____

(E) _____

_____

(E) _____

_____

■ **Figure 3.34 Right tibia and fibula in (a) anterior and (b) posterior views**

Q101-
104

IDENTIFY

**Tarsals
Metatarsals
Phalanges:**
  **proximal
  middle
  distal
  1st proximal phalanx**

Calcaneus
Cuboid
Intermediate cuneiform
Lateral cuneiform
Medial cuneiform
Navicular (scaphoid)
Talus (astragalus)

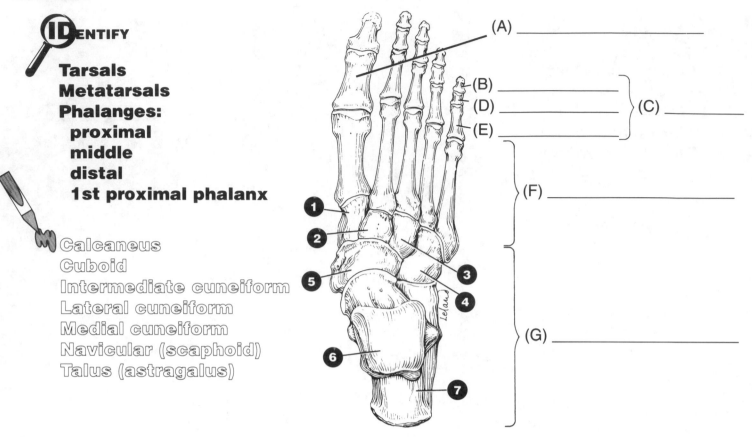

(A) _____

(B) _____
(D) _____   } (C) _____
(E) _____

(F) _____

(G) _____

■ Figure 3.35 Right ankle and foot, superior view

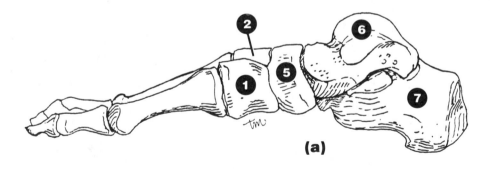

(a)

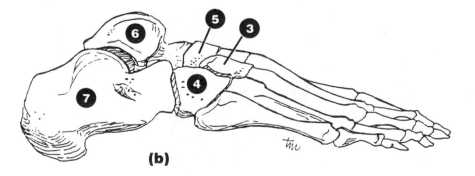

(b)

Q106-
110

■ Figure 3.36 Right ankle and foot in
(a) medial and (b) lateral views

MODULE 4

Arthrology

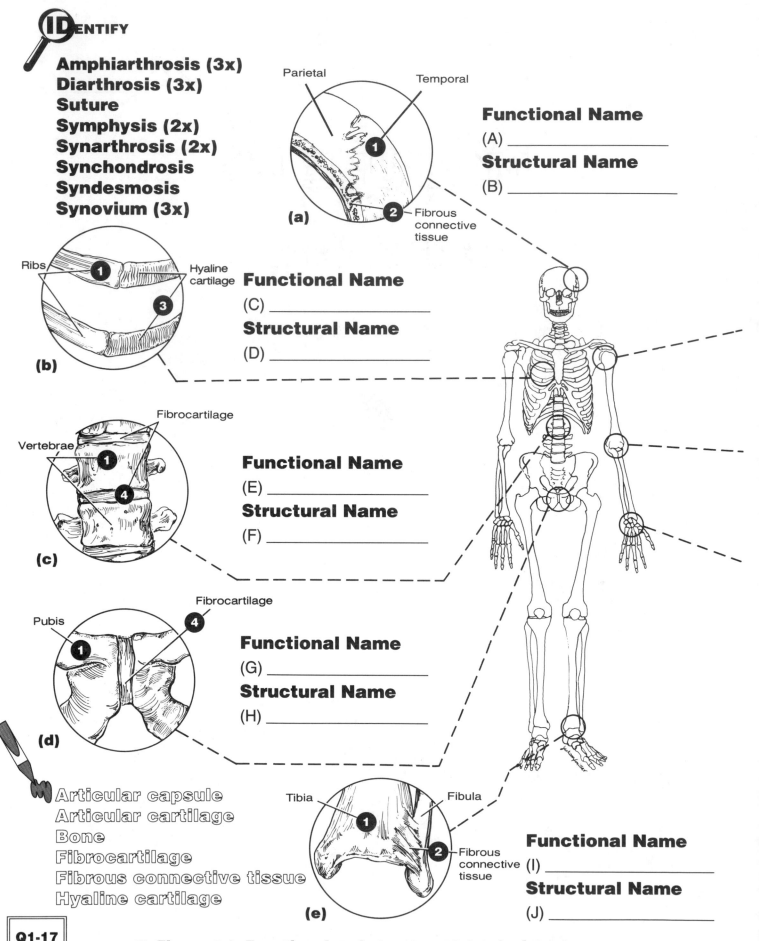

IDENTIFY

Amphiarthrosis (3x)
Diarthrosis (3x)
Suture
Symphysis (2x)
Synarthrosis (2x)
Synchondrosis
Syndesmosis
Synovium (3x)

(a)
Parietal
Temporal
1
2 Fibrous connective tissue

Functional Name
(A) _____
Structural Name
(B) _____

(b)
Ribs
1
Hyaline cartilage
3

Functional Name
(C) _____
Structural Name
(D) _____

(c)
Fibrocartilage
Vertebrae
1
4

Functional Name
(E) _____
Structural Name
(F) _____

(d)
Pubis
Fibrocartilage
4
1

Functional Name
(G) _____
Structural Name
(H) _____

Articular capsule
Articular cartilage
Bone
Fibrocartilage
Fibrous connective tissue
Hyaline cartilage

(e)
Tibia
Fibula
1
2 Fibrous connective tissue

Functional Name
(I) _____
Structural Name
(J) _____

Q1-17

■ Figure 4.1 Functional and structural joint designations

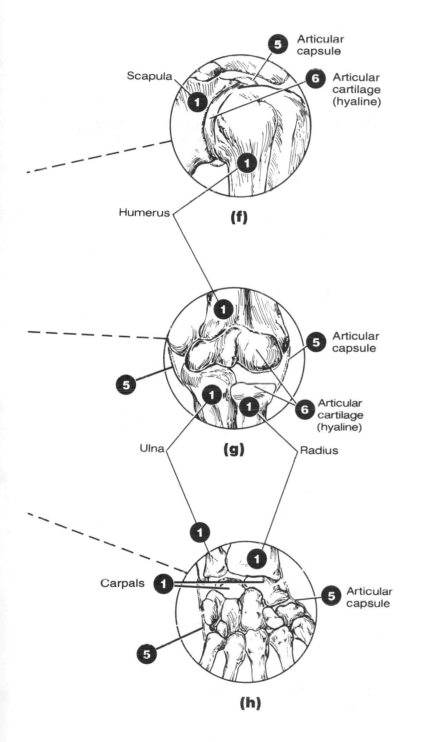

(f)

(g)

(h)

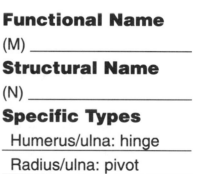

Functional Name

(K) _____

Structural Name

(L) _____

Specific Type

Spheroid (ball & socket)

Functional Name

(M) _____

Structural Name

(N) _____

Specific Types

Humerus/ulna: hinge

Radius/ulna: pivot

Functional Name

(O) _____

Structural Name

(P) _____

Specific Types

Radius/carpals: condyloid

Intercarpal: gliding

■ **Figure 4.1, continued Correlates with Table 4.1, page 50**

Q1-17

IDENTIFY

Amphiarthroses　　**Symphyses**　　　　**Syndesmoses**
Diarthroses　　　　**Synarthroses**　　　**Synovia**
Sutures　　　　　　**Synchondroses**

■ **Table 4.1 Summary of Human Joint Classification
Correlates with Figure 4.1**

Functional Name, Description	Structural Names	Structural Description	Specific Examples [See Fig. 4.1, (a)–(h)]
(A) _____ *Motion allowed: Not movable*	Fibrous	Joints bound by fibers	
	(B) _____	Interlocking bony edges bound by short fibers	(a) _____ _____
	(C) _____	Bones bound by long fibers	(b) _____ _____
(D) _____ *Motion allowed: Slightly movable*	Cartilaginous	Joints made of cartilage	
	(E) _____	Bones joined by fibrocartilage	(c),(d) _____ _____
	(F) _____	Bones joined by hyaline cartilage	(e) _____ _____
(G) _____ *Motion allowed: Freely movable*	(H) _____	Joints with articular capsule and cartilage, synovial membrane and fluid	
Nonaxial: Surfaces slide; no rotation	Gliding	Flat articulating surfaces	(f) _____ _____
Uniaxial: Rotate about one axis	Hinge and pivot joints	Complementary round or flat surfaces	(g) _____ _____
Biaxial: Rotate about two axes	Condyloid and saddle joints	Concave/convex surfaces	(h) _____ _____
Multiaxial: Rotate about three axes	Spheroid (ball and socket) joints	Spherical head fits cup-shaped cavity	(i) _____ _____

Q1-17

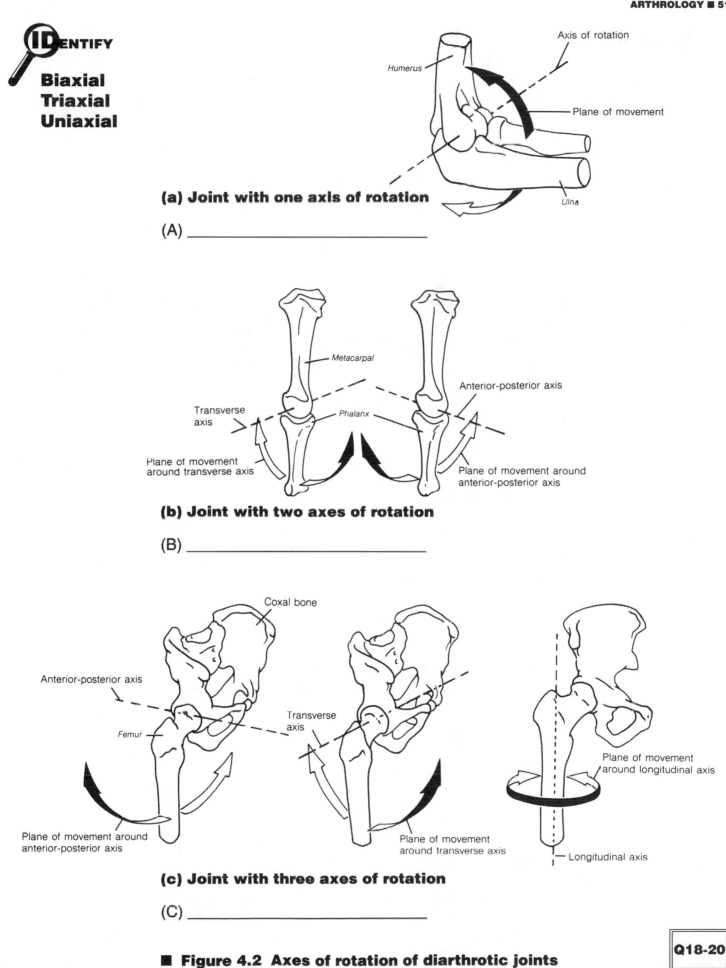

IDENTIFY

Biaxial
Triaxial
Uniaxial

(a) Joint with one axis of rotation

(A) _____

(b) Joint with two axes of rotation

(B) _____

(c) Joint with three axes of rotation

(C) _____

■ Figure 4.2 Axes of rotation of diarthrotic joints

Q18-20

IDENTIFY

Articular capsule
Articular cartilage
Meniscus
Synovial cavity with synovial fluid
Synovial membrane

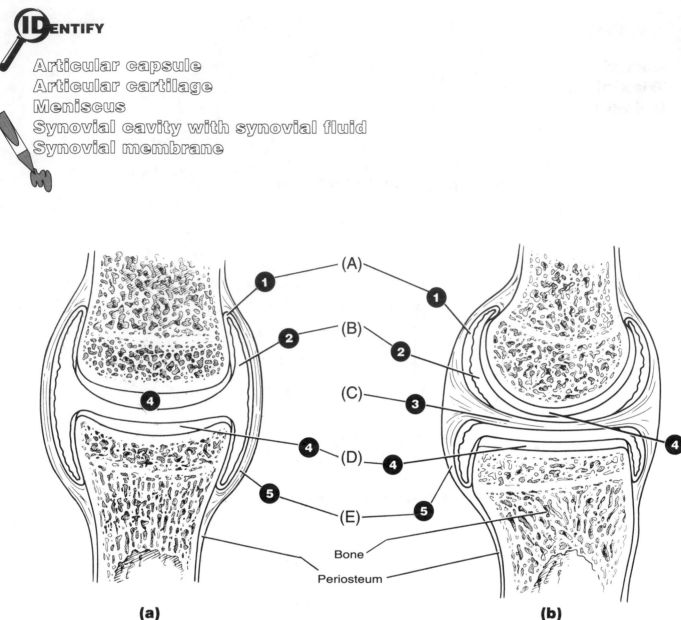

(a)   (b)

Bone
Periosteum

**■ Figure 4.3 Diarthrotic joint features with
(a) meniscus absent, (b) meniscus present**

Structure

(A) _____

(B) _____

(C) _____

(D) _____

(E) _____

IDENTIFY

Anterior cruciate ligament
Lateral collateral ligament
Lateral meniscus
Medial collateral ligament
Medial meniscus
Posterior cruciate ligament

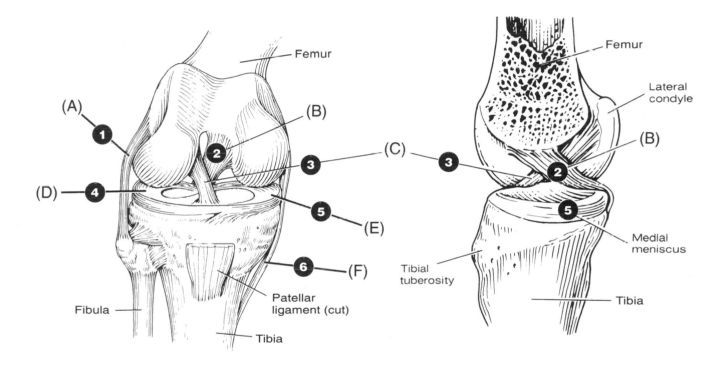

■ **Figure 4.4 Ligaments of the knee**

■ **Figure 4.5 Knee, medial view with femur sectioned**

Structure

(A) _____

(B) _____

(C) _____

(D) _____

(E) _____

(F) _____

Q26-32

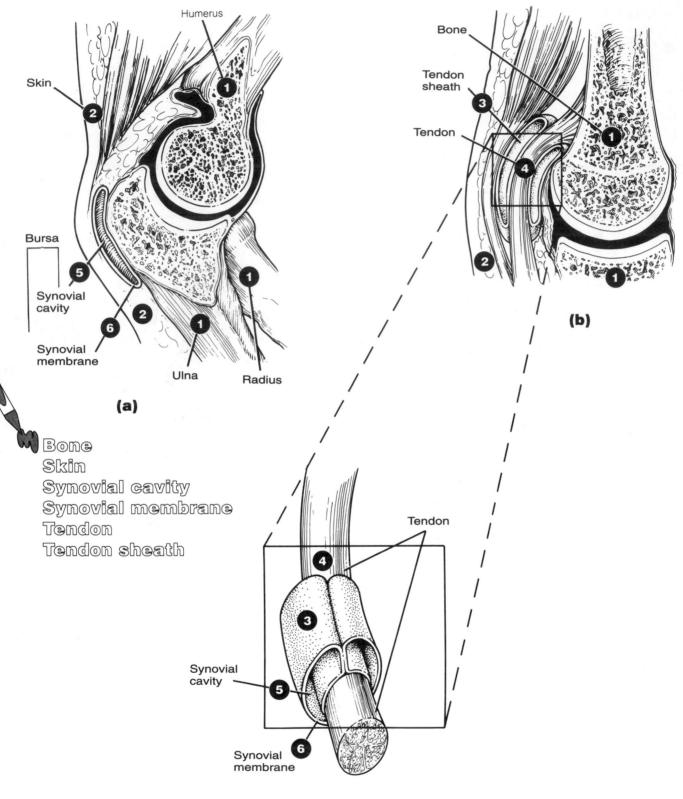

Bone
Skin
Synovial cavity
Synovial membrane
Tendon
Tendon sheath

■ Figure 4.6 Modified synovial membranes: (a) bursa, (b) tendon sheath, and (c) tendon sheath detail

Q33-34

MODULE 5

Muscle Actions

IDENTIFY

Insertion
Origin

In Figures 5.2–5.17:
All muscle actions
(Hint: Abduction,
Adduction, Depression,
Elevation, etc.)

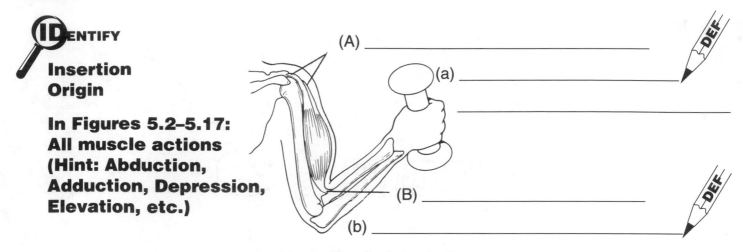

(A) _____

(a) _____

_____

(B) _____

(b) _____

■ Figure 5.1 Points of muscle attachment

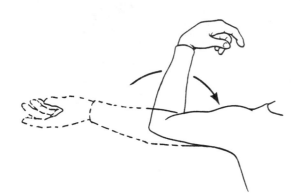

■ Figure 5.2

_____

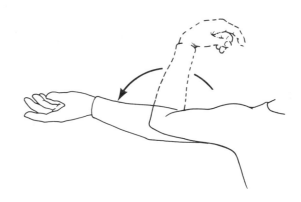

■ Figure 5.3

_____

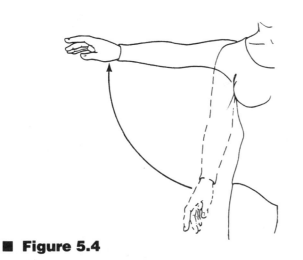

■ Figure 5.4

_____

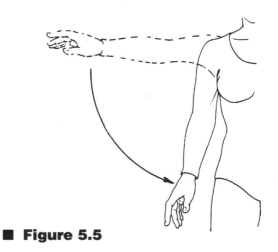

■ Figure 5.5

_____

Q1-6

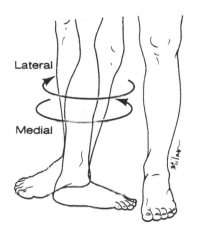

■ **Figure 5.6**

_____

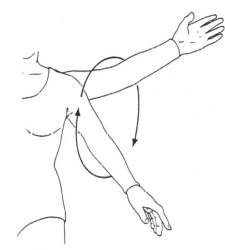

■ **Figure 5.7**

_____

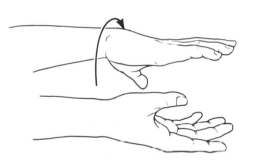

■ **Figure 5.8**

_____

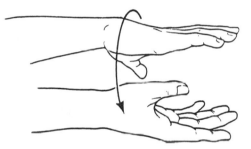

■ **Figure 5.9**

_____

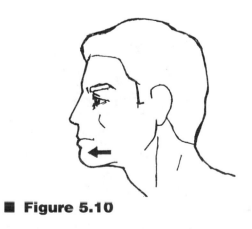

■ **Figure 5.10**

_____

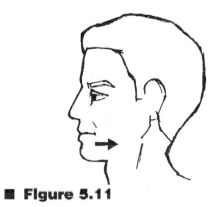

■ **Figure 5.11**

_____

Q7-10

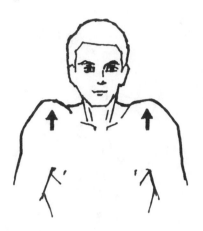

■ Figure 5.12

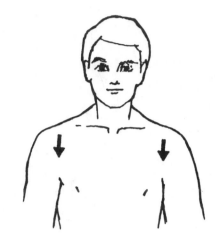

■ Figure 5.13

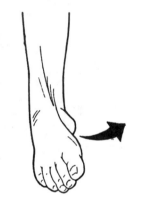

■ Figure 5.14

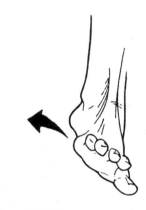

■ Figure 5.15

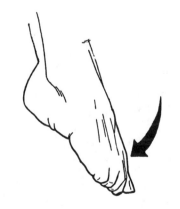

■ Figure 5.16

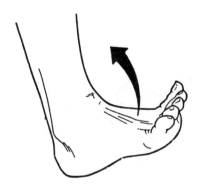

■ Figure 5.17

Q11-25

MODULE 6

Skeletal Muscles

Directory of Major Skeletal Muscles

The laboratory study of skeletal muscles is best done by dissection of a human cadaver or a nonhuman alternative, such as a cat. This module is not a substitute for dissection; however, as a supplement to your laboratory work, it will help you derive concise information on muscle attachments and functions.

For every figure in this module, you will be presented with a short list of muscles to identify and color. The colors you select can be used to highlight the muscle's name as given below the figure or on the adjacent page. In the space provided below the muscle's name you can write information regarding the muscle's ORIGIN, INSERTION and ACTION. A muscle's origin is its least movable point of attachment on the skeleton, while its insertion is its most movable attachment. The action of a muscle describes the motion it produces when the muscle shortens.

The major muscles of the body have been numbered for easy cross reference between the text and the answer key. They are grouped below by function. Numbers in parentheses (Q1-3, etc.) refer to the relevant questions generated by the accompanying software.

Muscles of the Face [Pages 64–65]
1. Orbicularis oculi (Q1-3)
2. Orbicularis oris (Q4-6)
3. Buccinator (Q7-9)
4. Platysma (Q10-12)

Muscles of Mastication [Page 66]
5. Temporalis (Q13-15)
6. Masseter (Q16-18)

Muscles of the Anterior Triangle of the Neck [Pages 67–69]
7. Stylohyoid (Q19-21)
8. Thyrohyoid (Q22-24)
9. Mylohyoid (Q28-30)
10. Digastric (Q25-27)
11. Sternocleidomastoid (Q37-39)
12. Sternohyoid (Q31-33)
13. Sternothyroid (Q34-36)

Muscles That Act on the Scapula [Pages 70–73]
14. (Clavo)trapezius (Q40-42)
15. Levator scapulae (Q52-54)
16. (Acromio)trapezius (Q43-45)
17. Rhomboideus (Q49-51)
18. (Spino)trapezius (Q46-48)
19. Pectoralis minor (Q55-57)
20. Serratus anterior (Q58-60)

Muscles That Act on the Arm (Humerus) [Pages 72–77]
21. Coracobrachialis (Q70-72)
22. Subscapularis (Q61-63)
23. (Acromio)deltoid (Q73-75)
24. (Clavo)deltoid (Q64-66)
25. Pectoralis major (Q67-69)
26. Supraspinatus (Q82-84)
27. (Spino)deltoid (Q76-78)
28. Infraspinatus (Q79-81)
29. Teres major (Q85-87)
30. Latissimus dorsi (Q88-90)

Muscles That Act on the Forearm (Radius and Ulna)[Pages 76–81]

31. Biceps brachii (Q91-93)
32. Brachialis (Q94-96)
33. Triceps brachii (Q97-99)
34. Pronator teres (Q103-105)
35. Brachioradialis (Q100-102)

Muscles That Act on the Wrist, Hand and Fingers [Pages 80–90]

36. Flexor carpi radialis (Q106-108)
37. Palmaris longus (Q109-111)
38. Flexor carpi ulnaris (Q112–114)
39. Ext. carpi radialis longus (Q115–117)
40. Flx. digitorum superficialis(Q118–120)
41. Flx. digitorum profundus (Q121–123)
42. Flexor pollicis longus (Q124–126)
43. Extensor digiti minimi (Q133–135)
44. Ext. carpi radialis brevis (Q127–129)
45. Ext. digitorum (Q130–132)
46. Extensor carpi ulnaris (Q136–138)
47. Supinator (Q139–141)
48. Abductor pollicis longus (Q145–147)
49. Extensor pollicis longus (Q142–143)
50. Abductor pollicis brevis (Q148–150)
51. Adductor pollicis (Q151–153)

Deep Muscles of the Thorax (Respiratory Muscles) [Page 91]

52. External intercostals (Q154-156)
53. Internal intercostals (Q157-159)

Muscles That Move the Vertebral Column [Pages 92–93]

54. Erector spinae (Q160-162)

Muscles of the Abdominal Wall [Pages 94–95]

55. Ext. abdominal oblique (Q163-165)
56. Rectus abdominis (Q172-174)
57. Transversus abdominis (Q169-171)
58. Int. abdominal oblique (Q166-168)

Muscles That Act on the Thigh (Femur) [Pages 96–101]

59. Psoas major (Q175-177)
60. Iliacus (Q178-180)
61. Pectineus (Q181-183)
62. Gluteus medius (Q187-189)
63. Gluteus maximus (Q184-186)
64. Tensor fasciae latae (Q190-192)
65. Adductor brevis (Q199-201)
66. Adductor longus (Q196-198)
67. Adductor magnus (Q193-195)

Muscles That Act on the Leg (Muscles of the Thigh) [Pages 102–107]

68. Sartorius (Q205-207)
69. Gracilis (Q202-204)
70. Rectus femoris (Q208-210)
71. Vastus lateralis (Q211-213)
72. Vastus medialis (Q214-216)
73. Vastus intermedius (Q217-219)
74. Semitendinosus (Q223-225)
75. Biceps femoris (Q220-222)
76. Semimembranosus (Q226-228)

Muscles That Act on the Foot and Toes (Muscles of the Leg) [Pages 108–109]

77. Tibialis anterior (Q229-231)
78. Ext. digitorum longus (Q232-234)
79. Peroneus longus (Q235-237)
80. Peroneus brevis (Q238-240)
81. Peroneus tertius (Q241-243)
82. Extensor hallucis longus (Q244-246)
83. Gastrocnemius (Q247-249)
84. Soleus (Q250-252)
85. Tibialis posterior (Q256-258)
86. Flexor hallucis longus (Q253-255)
87. Flexor digitorum longus (Q259-261)
88. Flexor digitorum brevis (Q262-264)

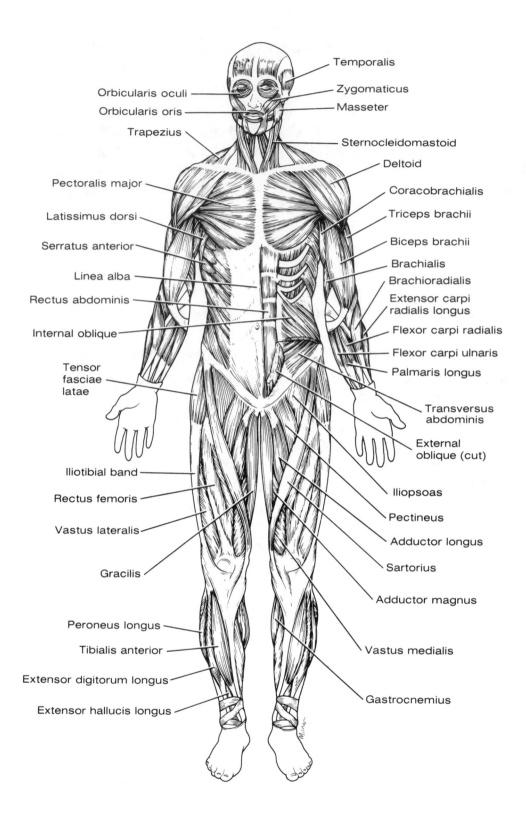

■ Figure 6.1 General overview of major muscles, anterior view

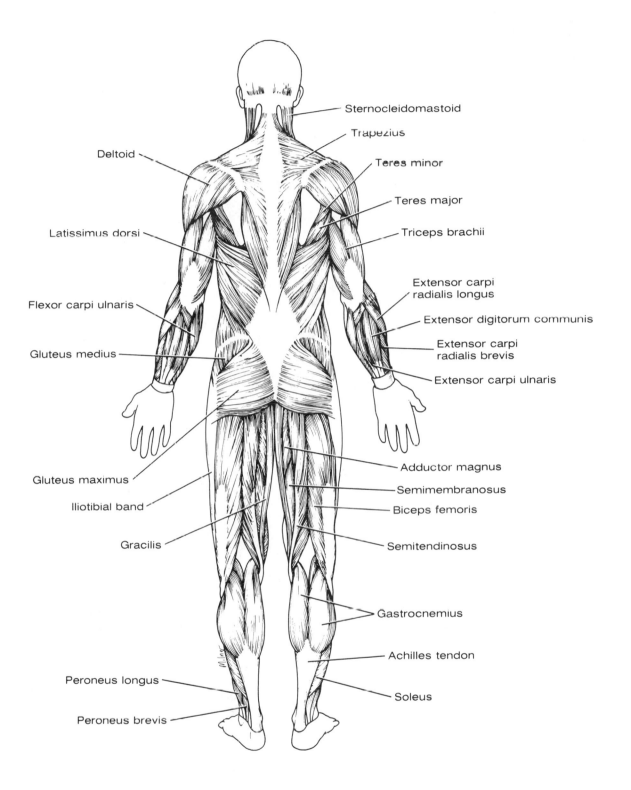

Sternocleidomastoid

Trapezius

Teres minor

Teres major

Triceps brachii

Extensor carpi radialis longus

Extensor digitorum communis

Extensor carpi radialis brevis

Extensor carpi ulnaris

Deltoid

Latissimus dorsi

Flexor carpi ulnaris

Gluteus medius

Adductor magnus

Semimembranosus

Biceps femoris

Semitendinosus

Gluteus maximus

Iliotibial band

Gracilis

Gastrocnemius

Achilles tendon

Peroneus longus

Soleus

Peroneus brevis

■ **Figure 6.2 General overview of major muscles, posterior view**

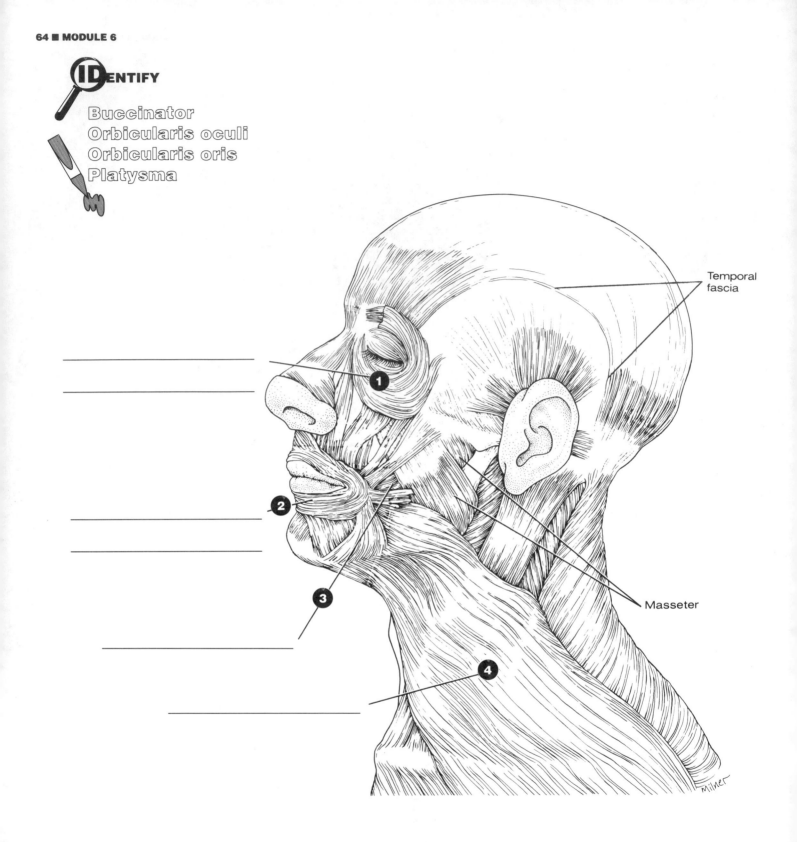

IDENTIFY

Buccinator
Orbicularis oculi
Orbicularis oris
Platysma

Temporal
fascia

Masseter

■ **Figure 6.3 Muscles of the face. The temporalis
muscle lies deep to the temporal fascia.**

Muscles of the Face

Buccinator

Origin: _____

Insertion: _____

Action: _____

Orbicularis oculi

Origin: _____

Insertion: _____

Action: _____

Orbicularis oris

Origin: _____

Insertion: _____

Action: _____

Platysma

Origin: _____

Insertion: _____

Action: _____

Q1-12

IDENTIFY

Masseter
Temporalis

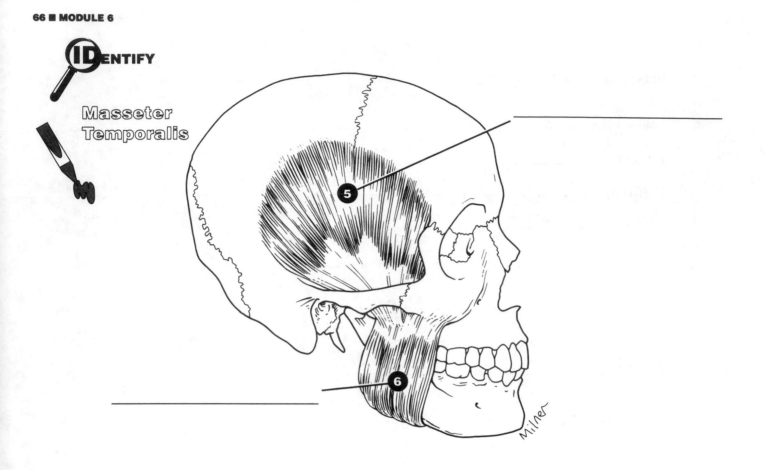

■ **Figure 6.4 Major muscles of mastication**

Masseter

Origin:_____

Insertion:_____

Action:_____

Temporalis

Origin:_____

Insertion:_____

Action:_____

Q13-18

IDENTIFY

Stylohyoid
Thyrohyoid

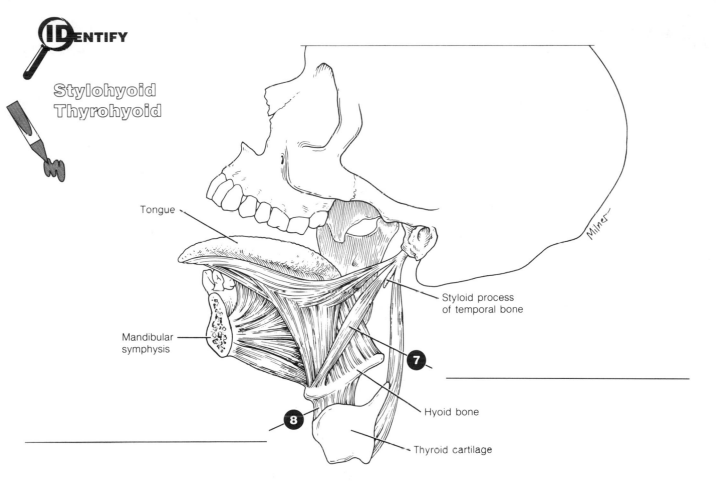

Tongue

Styloid process
of temporal bone

Mandibular
symphysis

7

Hyoid bone

8

Thyroid cartilage

Milner

■ **Figure 6.5 Tongue and hyoid musculature**

Muscles of the Anterior Triangle of the Neck

Stylohyoid

Origin:_____

Insertion:_____

Action:_____

Thyrohyoid

Origin:_____

Insertion:_____

Action:_____

IDENTIFY

Digastric
Mylohyoid
Sternohyoid
Sternothyroid
Sternocleidomastoid

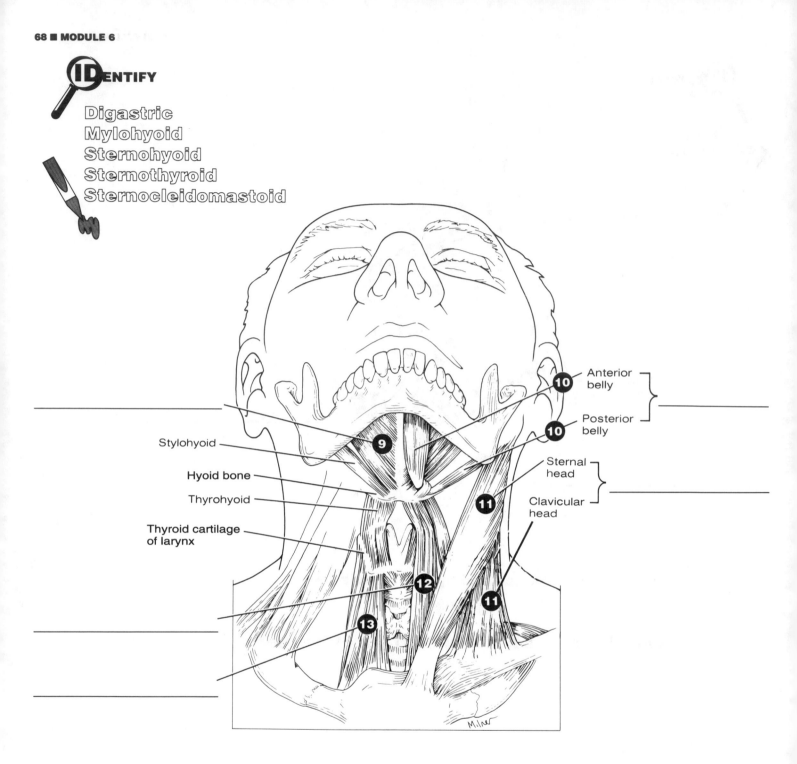

Anterior belly

Posterior belly

Sternal head

Clavicular head

Stylohyoid

Hyoid bone

Thyrohyoid

Thyroid cartilage of larynx

Milner

■ **Figure 6.6 Muscles of the anterior neck. Sternocleidomastoid and digastric have been removed on the right side.**

Muscles of the Anterior Triangle of the Neck

Digastric

Origin: _____

Insertion: _____

Action: _____

Mylohyoid

Origin: _____

Insertion: _____

Action: _____

Sternocleidomastoid

Origin: _____

Insertion: _____

Action: _____

Sternohyoid

Origin: _____

Insertion: _____

Action: _____

Sternothyroid

Origin: _____

Insertion: _____

Action: _____

Q25-39

ID ENTIFY

(Acromio)trapezius
(Clavo)trapezius
Levator scapulae
Rhomboideus
(Spino)trapezius

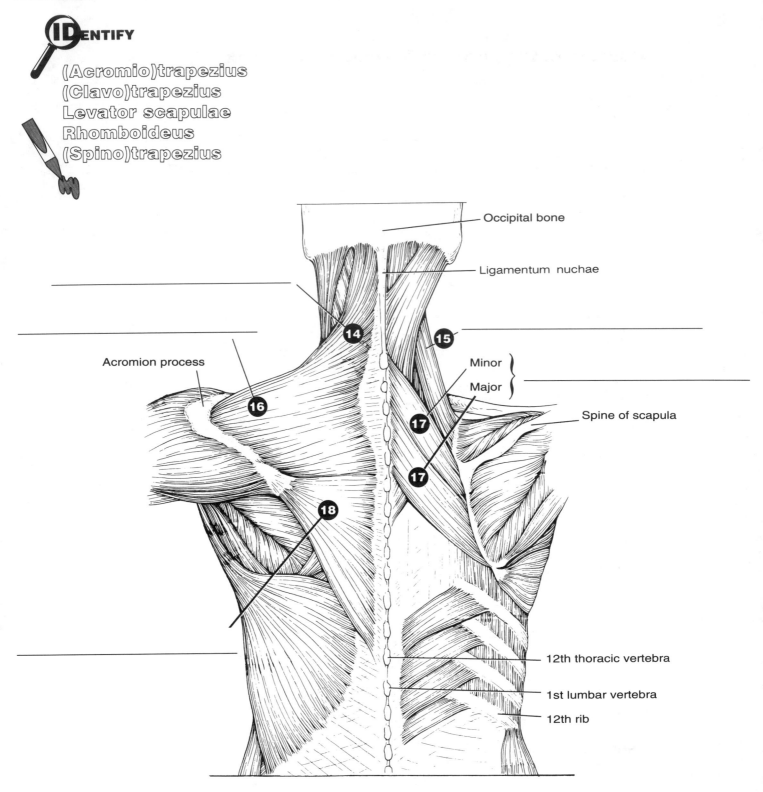

■ Figure 6.7 Muscles of the back. Superficial group
is on the left; deep group is on the right.

Muscles That Act on the Scapula

(Acromio)trapezius

Origin: _____

Insertion: _____

Action: _____

(Clavo)trapezius

Origin: _____

Insertion: _____

Action: _____

Levator scapulae

Origin: _____

Insertion: _____

Action: _____

Rhomboideus

Origin: _____

Insertion: _____

Action: _____

(Spino)trapezius

Origin: _____

Insertion: _____

Action: _____

IDENTIFY

Coracobrachialis
Pectoralis minor
Serratus anterior
Subscapularis

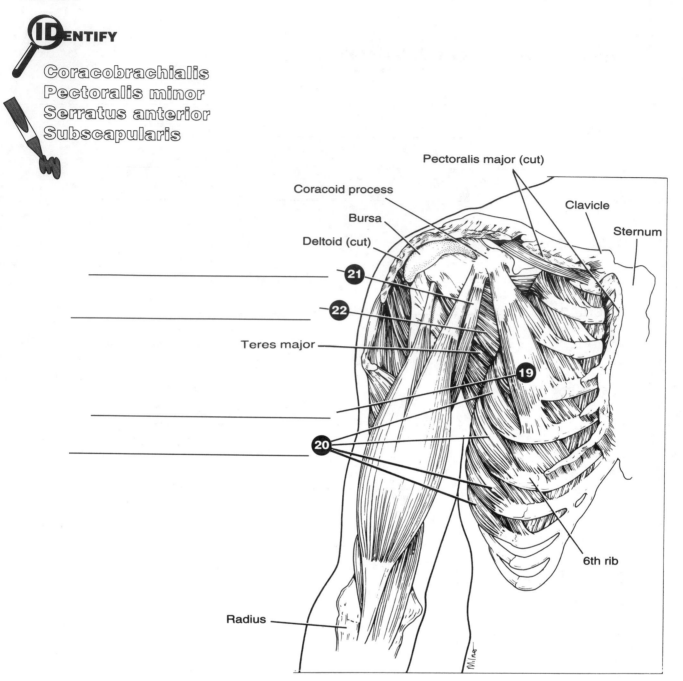

Pectoralis major (cut)

Coracoid process

Clavicle

Bursa

Sternum

Deltoid (cut)

21

22

Teres major

19

20

6th rib

Radius

■ **Figure 6.8 Deep muscles of the anterior chest**

Muscles That Act on the Scapula

Pectoralis minor

Origin: _____

Insertion: _____

Action: _____

Serratus anterior

Origin: _____

Insertion: _____

Action: _____

Muscles That Act on the Arm (Humerus)

Coracobrachialis

Origin: _____

Insertion: _____

Action: _____

Subscapularis

Origin: _____

Insertion: _____

Action: _____

Q55-63
Q70-72

IDENTIFY

(Acromio)deltoid
(Clavo)deltoid
Pectoralis major

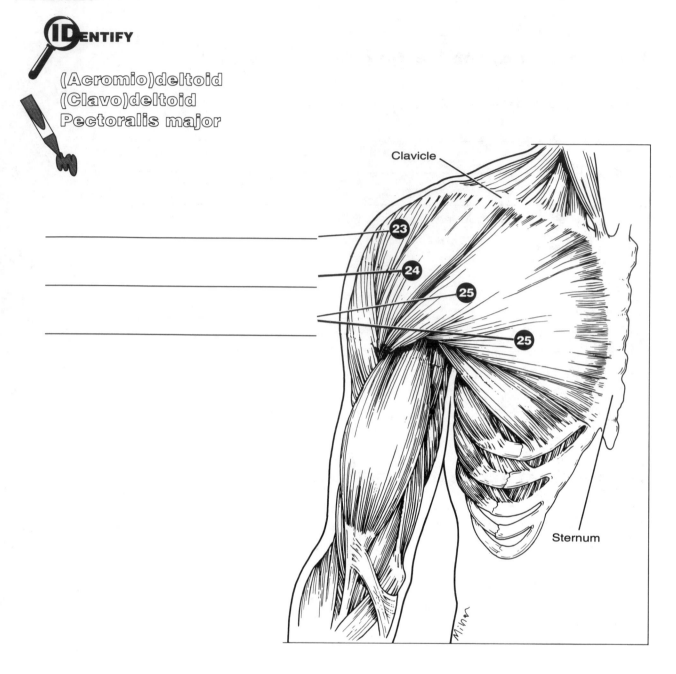

■ Figure 6.9 Superficial muscles of the anterior chest

Muscles That Act on the Arm (Humerus)

(Acromio)deltoid

Origin: _____

Insertion: _____

Action: _____

(Clavo)deltoid

Origin: _____

Insertion: _____

Action: _____

Pectoralis major

Origin: _____

Insertion: _____

Action: _____

Q64-69
Q73-75

IDENTIFY

Infraspinatus
Latissimus dorsi
(Spino)deltoid
Supraspinatus
Teres major

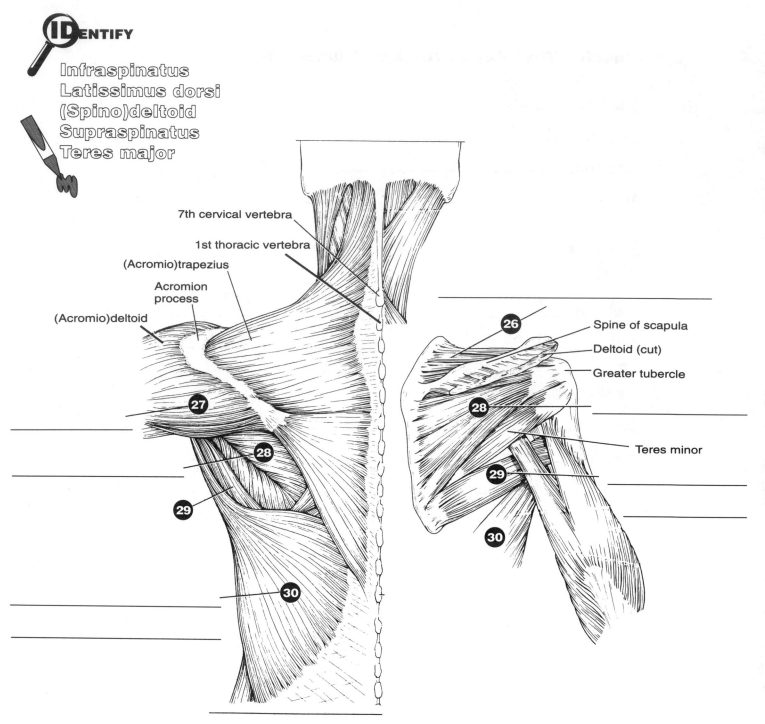

7th cervical vertebra

1st thoracic vertebra

(Acromio)trapezius

Acromion
process

(Acromio)deltoid

26

Spine of scapula

Deltoid (cut)

Greater tubercle

27

28

Teres minor

28

29

29

30

30

■ Figure 6.10 Muscles of the shoulder and back. Superficial
group is on the left; deep group is on the right.

Muscles That Act on the Arm (Humerus)

Infraspinatus

Origin: _____

Insertion: _____

Action: _____

Latissimus dorsi

Origin: _____

Insertion: _____

Action: _____

(Spino)deltoid

Origin: _____

Insertion: _____

Action: _____

Supraspinatus

Origin: _____

Insertion: _____

Action: _____

Teres major

Origin: _____

Insertion: _____

Action: _____

Q76-90

IDENTIFY

Biceps brachii
Brachialis

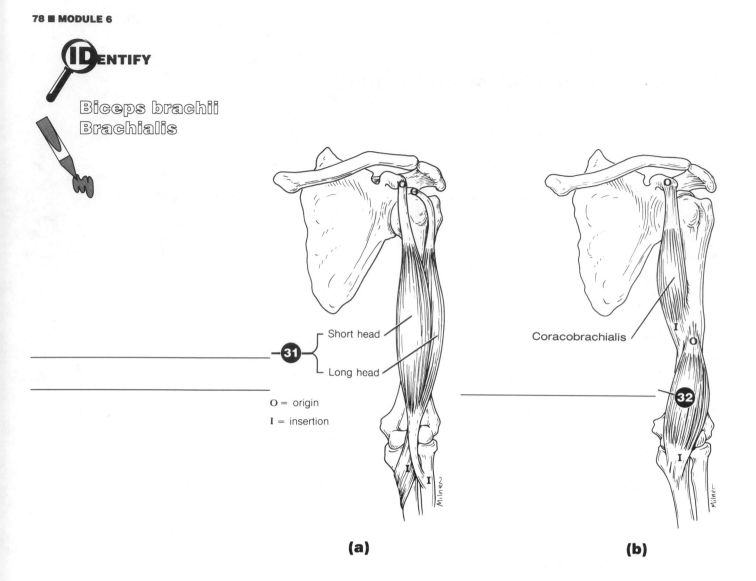

Short head
31
Long head

O = origin
I = insertion

Coracobrachialis

I
O
32

I

(a)   (b)

■ **Figure 6.11 Anterior views of the arm: (a) superficial and (b) deep**

Muscles That Act on the Forearm (Radius and Ulna)

Biceps brachii

Origins: _____

_____

Insertion: _____

Action: _____

Brachialis

Origin: _____

Insertion: _____

Q91-96

Action: _____

IDENTIFY

Triceps brachii

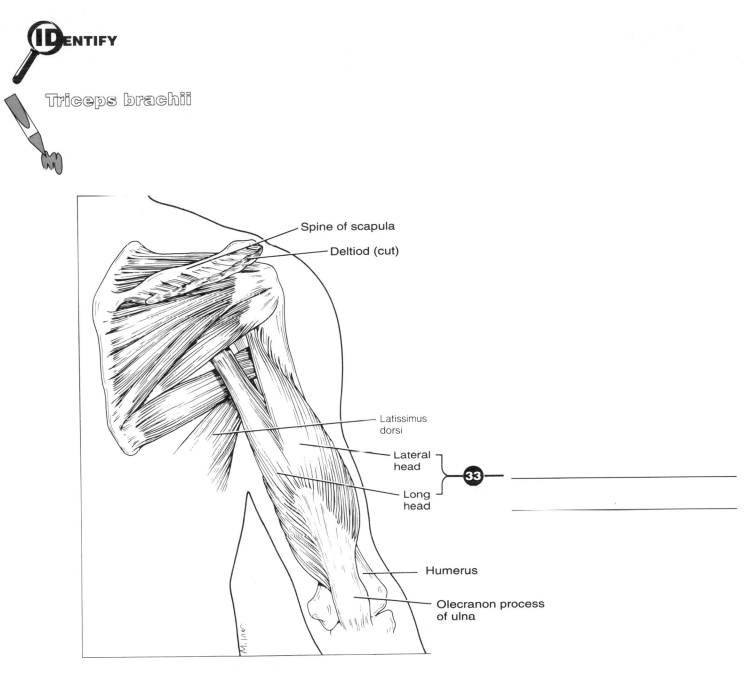

■ **Figure 6.12 Superficial view of posterior arm and shoulder**

Muscles That Act on the Forearm (Radius and Ulna)

Triceps brachii

Origins: _____

_____

Insertion: _____

Action: _____

Q97-99

ID ENTIFY

Brachioradialis
Flexor carpi radialis
Flexor carpi ulnaris
Palmaris longus
Pronator teres

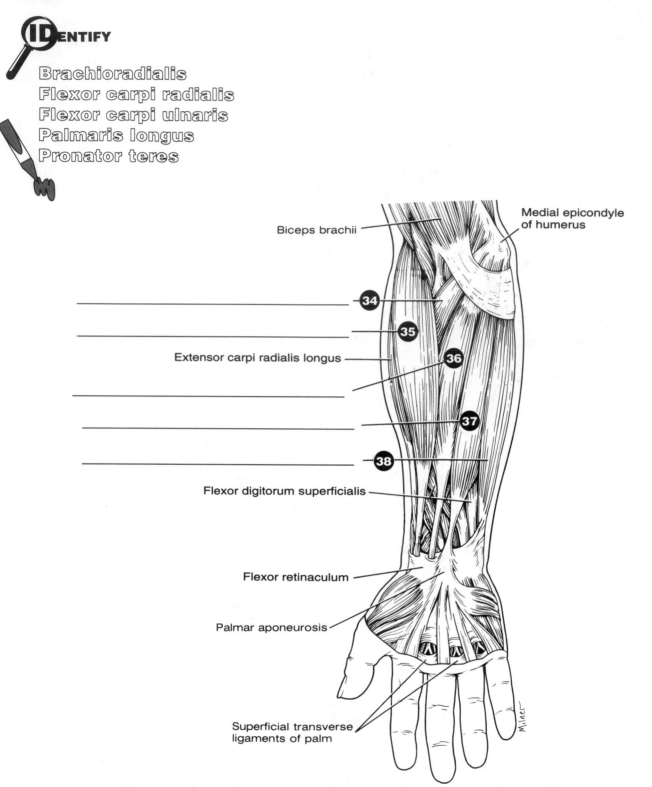

Biceps brachii

Medial epicondyle
of humerus

34

35

Extensor carpi radialis longus

36

37

38

Flexor digitorum superficialis

Flexor retinaculum

Palmar aponeurosis

Superficial transverse
ligaments of palm

■ Figure 6.13 Muscles of the anterior forearm, superficial group

Muscles That Act on the Forearm (Radius and Ulna)

Brachioradialis

Origin: _____

Insertion: _____

Action: _____

Pronator teres

Origin: _____

Insertion: _____

Action: _____

Muscles That Act on the Wrist and Hand

Flexor carpi radialis

Origin: _____

Insertion: _____

Action: _____

Flexor carpi ulnaris

Origin: _____

Insertion: _____

Action: _____

Palmaris longus

Origin: _____

Insertion: _____

Action: _____

Extensor carpi radialis longus
Flexor digitorum superficialis

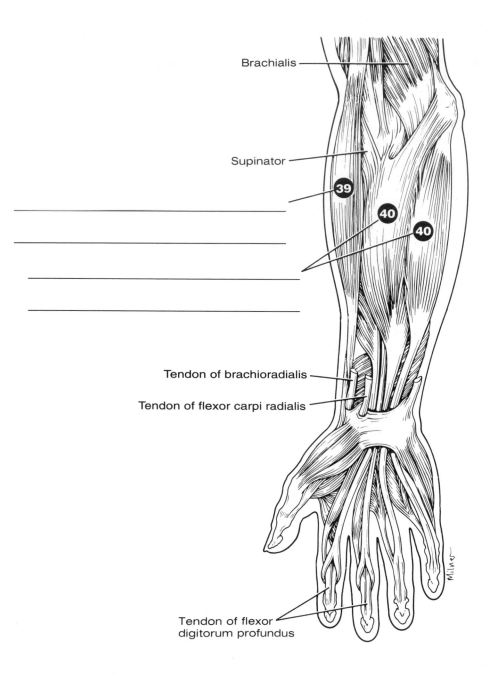

Brachialis

Supinator

39

40

40

Tendon of brachioradialis

Tendon of flexor carpi radialis

Tendon of flexor
digitorum profundus

■ **Figure 6.14 Muscles of the anterior forearm, middle group**

Muscles That Act on the Wrist, Hand, and Fingers

Extensor carpi radialis longus

Origin: _____ _____

Insertion: _____ _____

Action: _____

Flexor digitorum superficialis

Origin: _____

Insertion: _____

Action: _____

IDENTIFY

Flexor digitorum profundus
Flexor pollicis longus

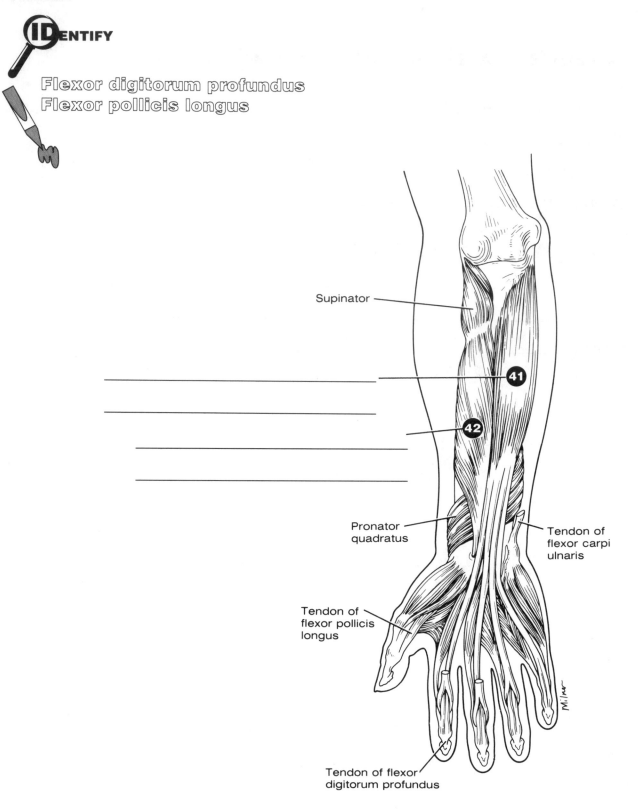

Supinator

Pronator
quadratus

Tendon of
flexor carpi
ulnaris

Tendon of
flexor pollicis
longus

Tendon of flexor
digitorum profundus

■ **Figure 6.15 Muscles of the anterior forearm, deep group**

Muscles That Act on the Wrist, Hand, and Fingers

Flexor digitorum profundus

Origin: _____

Insertion: _____

Action: _____

Flexor pollicis longus

Origin: _____

Insertion: _____

Action: _____

ID ENTIFY

Extensor carpi radialis brevis
Extensor carpi ulnaris
Extensor digiti minimi
Extensor digitorum

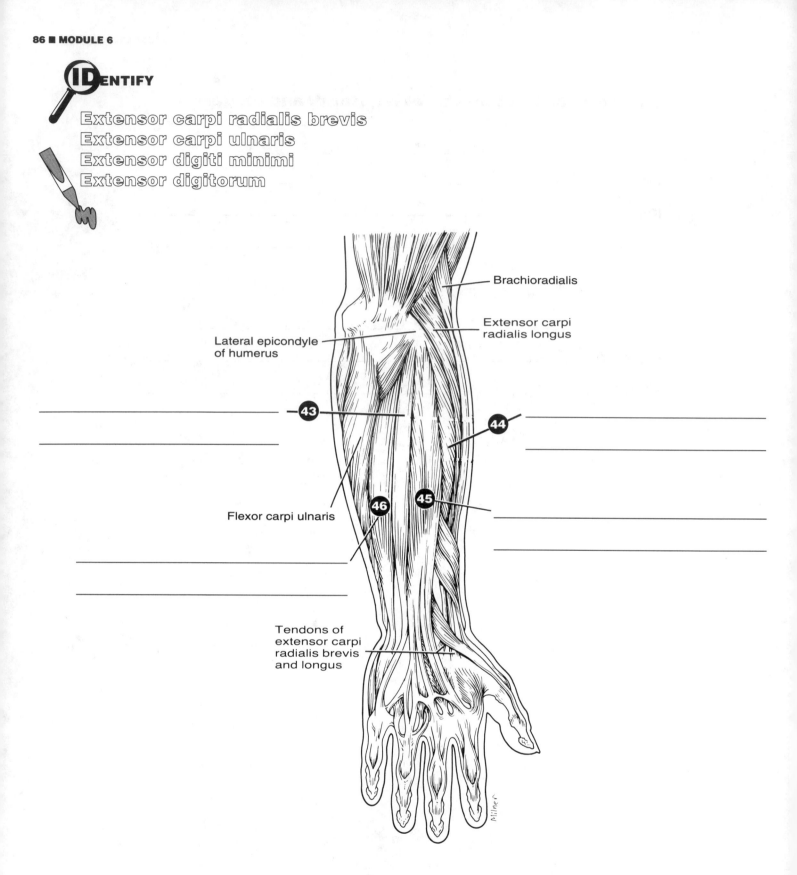

Brachioradialis

Extensor carpi
radialis longus

Lateral epicondyle
of humerus

43

44

Flexor carpi ulnaris

46

45

Tendons of
extensor carpi
radialis brevis
and longus

Milner

■ Figure 6.16 Muscles of the posterior forearm, superficial group

Muscles That Act on the Wrist, Hand, and Fingers

Extensor carpi radialis brevis

Origin:

Insertion:

Action:

Extensor carpi ulnaris

Origin:

Insertion:

Action:

Extensor digiti minimi

Origin:

Insertion:

Action:

Extensor digitorum

Origin:

Insertion:

Action:

Abductor pollicis longus
Extensor pollicis longus
Supinator

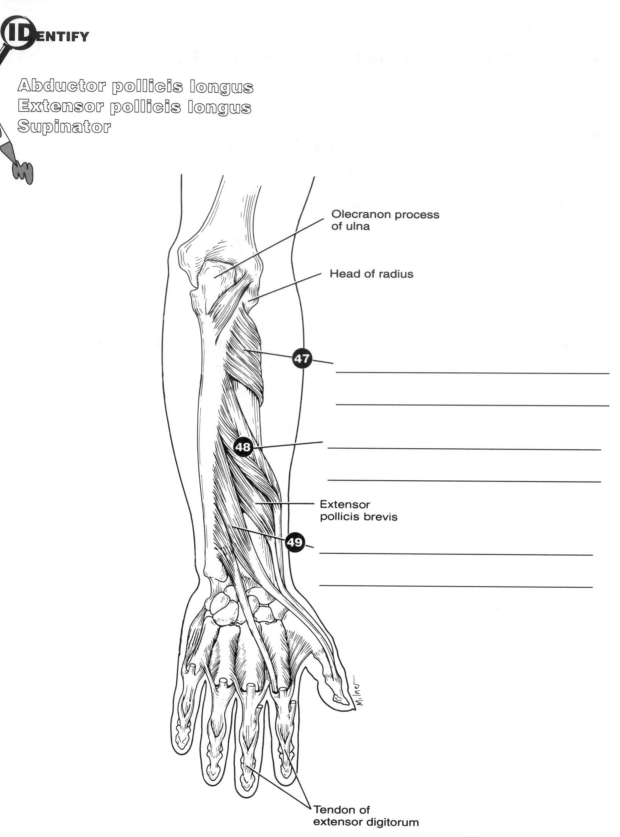

Olecranon process
of ulna

Head of radius

47

48

Extensor
pollicis brevis

49

Tendon of
extensor digitorum

■ **Figure 6.17 Muscles of the posterior forearm, deep group**

Muscles That Act on the Wrist, Hand, and Fingers

Abductor pollicis longus

Origin: _____

Insertion: _____

Action: _____

Extensor pollicis longus

Origin: _____

Insertion: _____

Action: _____

Supinator

Origin: _____

Insertion: _____

Action: _____

IDENTIFY

Abductor pollicis brevis
Adductor pollicis

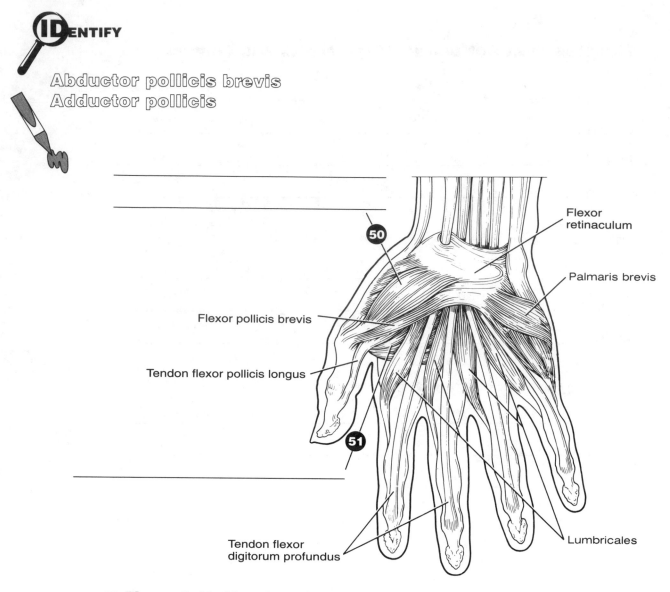

Flexor
retinaculum

Palmaris brevis

50

Flexor pollicis brevis

Tendon flexor pollicis longus

51

Lumbricales

Tendon flexor
digitorum profundus

■ **Figure 6.18 Muscles of the hand, palmar group**

Muscles That Act on the Wrist, Hand, and Fingers

Abductor pollicis brevis

Origin:

Insertion:

Action:

Adductor pollicis

Origin:

Insertion:

Action:

IDENTIFY

External intercostals
Internal intercostals

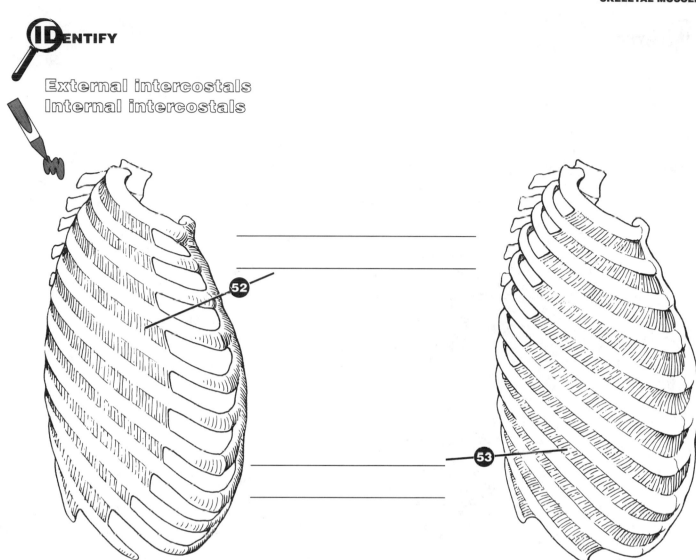

■ **Figure 6.19 Muscles of the thorax, deep group**

Respiratory Muscles

External intercostals

Origin:_____

Insertion:_____

Action:_____

Internal intercostals

Origin:_____

Insertion:_____

Action:_____

IDENTIFY

Erector spinae

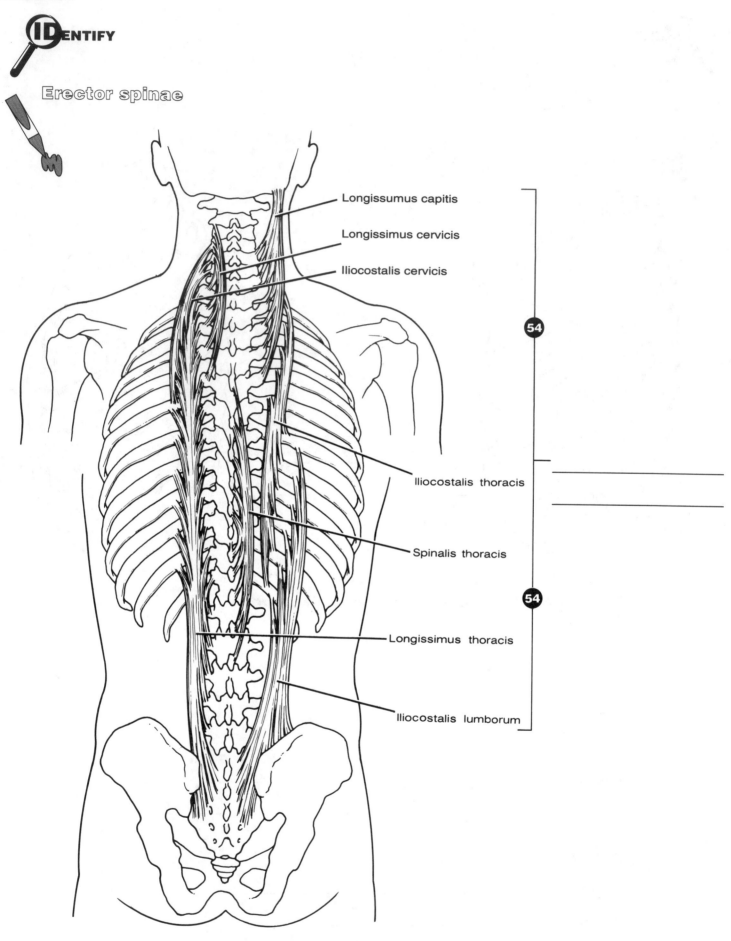

Longissumus capitis

Longissimus cervicis

Iliocostalis cervicis

54

Iliocostalis thoracis

Spinalis thoracis

54

Longissimus thoracis

Iliocostalis lumborum

■ **Figure 6.20 Muscles of the back, deep group**

Muscles That Move the Vertebral column

Erector spinae

Origins: _____

_____

Insertion: _____

Action: _____

■ Table 6.1 Specific Components of Erector Spinae

Muscle	Origin	Insertion
Iliocostalis lumborum thoracis cervicis	Crest of the sacrum; spinous processes of lumbar and lower thoracic vertebrae; iliac crests; angles of the ribs	Angles of the ribs; transverse processes of the cervical vertebrae
Longissimus thoracis cervicis capitis	Transverse processes of the lumbar, thoracic, and lower cervical vertebrae	Transverse processes of the vertebra above the vertebra of origin, and the mastoid process of the temporal bone (capitis)
Spinalis thoracis cervicis	Spinous process of the upper lumbar, lower thoracic, and seventh cervical vertebrae	Spinous processes of the upper thoracic and the cervical vertebrae

IDENTIFY

External abdominal oblique
Internal abdominal oblique
Rectus abdominis
Transversus abdominis

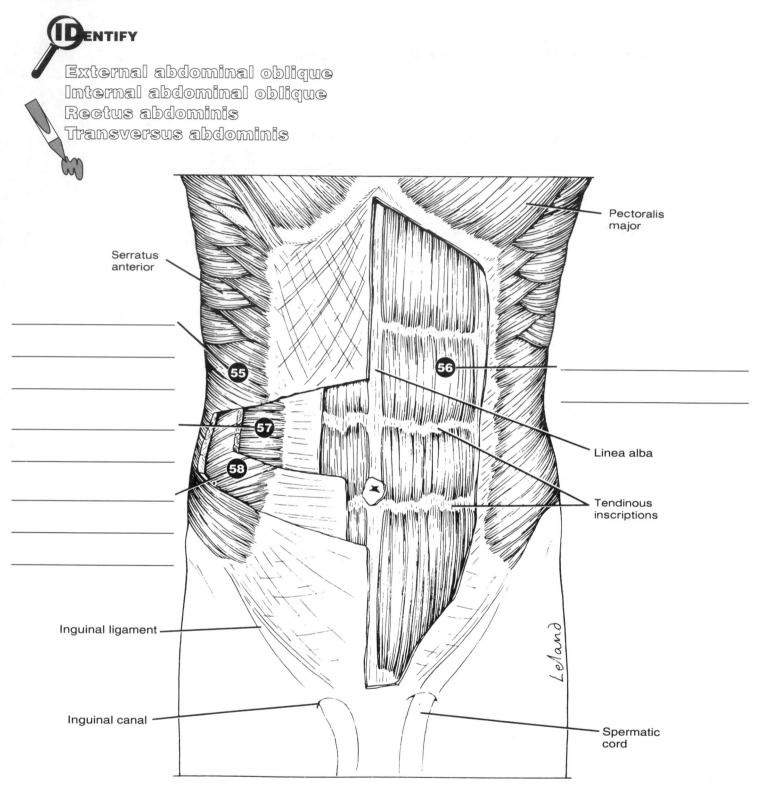

Pectoralis major

Serratus anterior

55

56

57

58

Linea alba

Tendinous inscriptions

Inguinal ligament

Inguinal canal

Spermatic cord

LeLand

■ **Figure 6.21 Abdominal muscles. Deep group is on the specimen's right; superficial group is on the left.**

Muscles of the Abdominal Wall

External abdominal oblique

Origin: _____

Insertion: _____

Action: _____

Internal abdominal oblique

Origin: _____

Insertion: _____

Action: _____

Rectus abdominis

Origin: _____

Insertion: _____

Action: _____

Transversus abdominis

Origin: _____

Insertion: _____

Action: _____

IDENTIFY

Iliacus
Pectineus
Psoas major

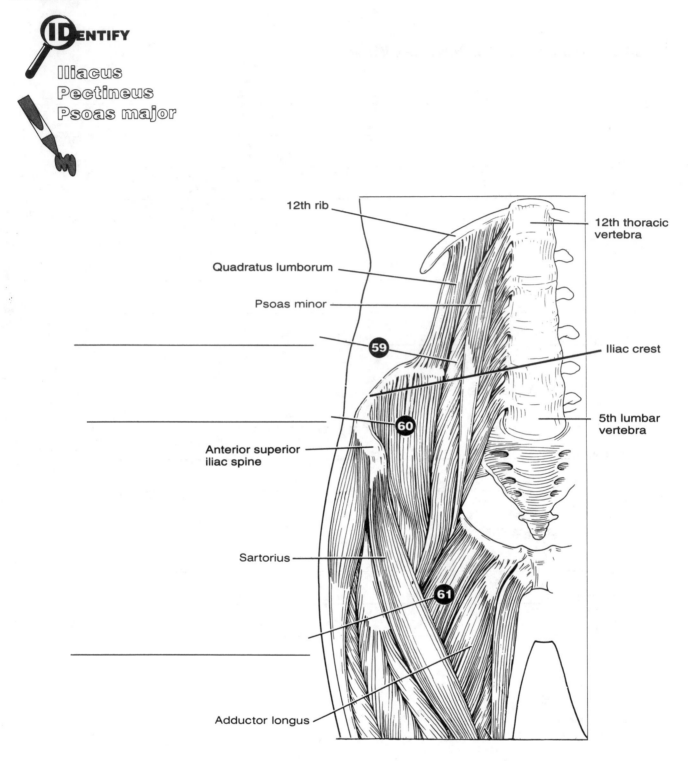

12th rib

Quadratus lumborum

Psoas minor

59

60

Anterior superior
iliac spine

Sartorius

61

Adductor longus

12th thoracic
vertebra

Iliac crest

5th lumbar
vertebra

■ **Figure 6.22 Muscles of the lumbar vertebrae and
pelvis that attach to the femur, anterior group**

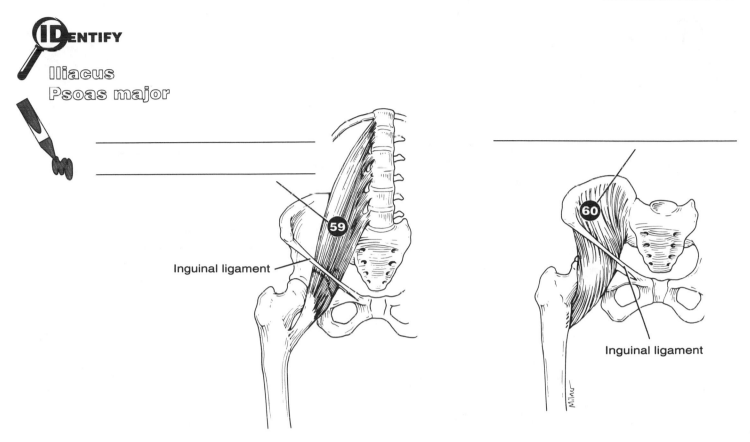

ID**ENTIFY**

Iliacus

Psoas major

Inguinal ligament

Inguinal ligament

■ **Figure 6.23 Exposure of psoas major and iliacus muscles**

Muscles That Act on the Thigh (Femur)

Iliacus

Origin: _____

Insertion: _____

Action: _____

Pectineus

Origin: _____

Insertion: _____

Action: _____

Psoas major

Origin: _____

Insertion: _____

Action: _____

IDENTIFY

Gluteus maximus
Gluteus medius

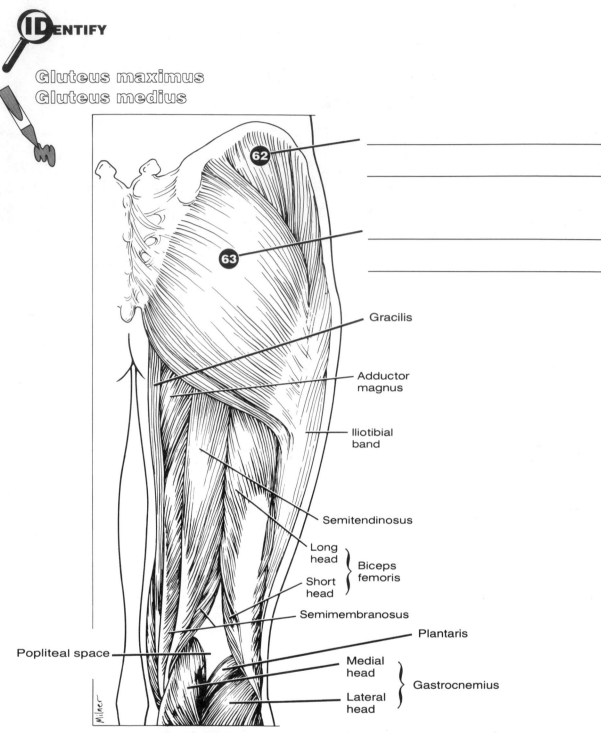

62

63

Gracilis

Adductor
magnus

Iliotibial
band

Semitendinosus

Long
head ⎱
      ⎰ Biceps
Short   femoris
head ⎰

Semimembranosus

Popliteal space

Plantaris

Medial
head ⎱
      ⎰ Gastrocnemius
Lateral
head ⎰

Milner

■ Figure 6.24 Superficial muscles of the posterior hip and thigh

Muscles That Act on the Thigh (Femur)

Gluteus maximus

Origin: _____

Insertion: _____

Q184-
186

Action: _____

IDENTIFY

Gluteus maximus
Gluteus medius

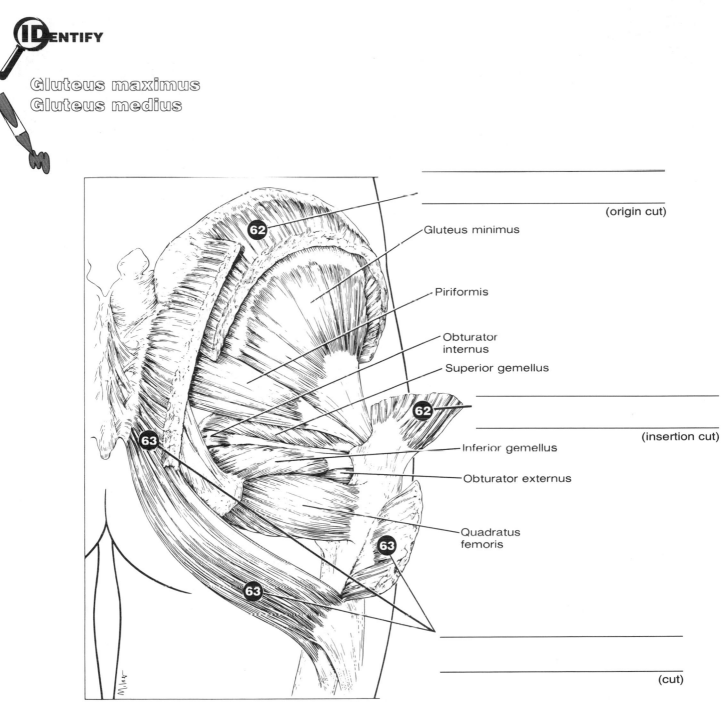

_____

_____
(origin cut)

Gluteus minimus

Piriformis

Obturator
internus

Superior gemellus

_____

_____
(insertion cut)

Inferior gemellus

Obturator externus

Quadratus
femoris

_____

_____
(cut)

■ **Figure 6.25 Muscles of the posterior hip, deep group**

Muscles That Act on the Thigh (Femur)

Gluteus medius

Origin:_____

Insertion:_____

Action:_____

Q187-
189

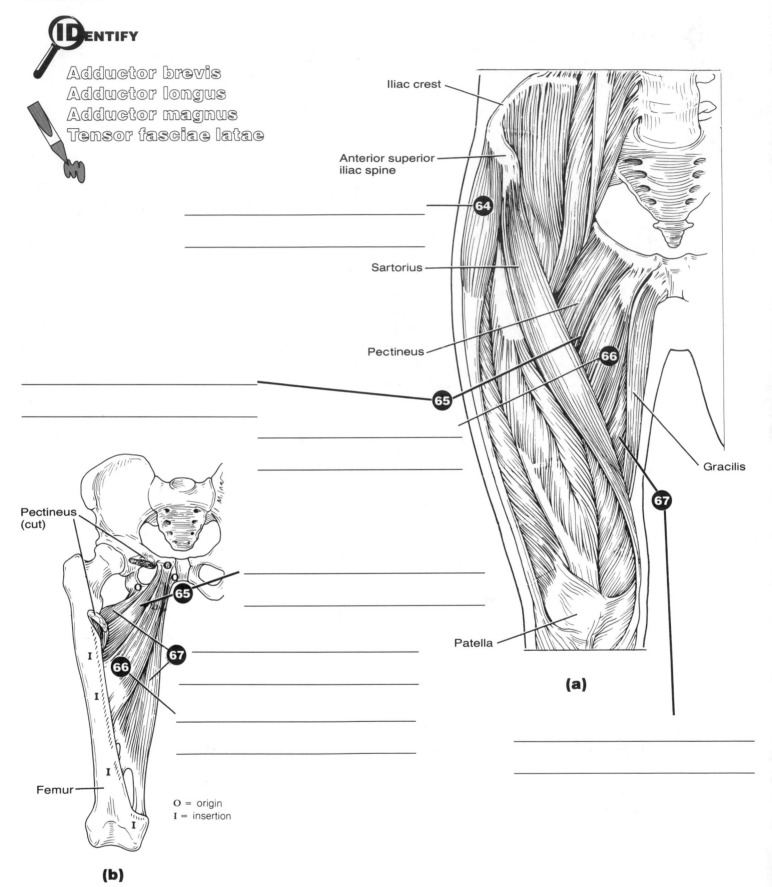

IDENTIFY

Adductor brevis
Adductor longus
Adductor magnus
Tensor fasciae latae

Iliac crest

Anterior superior
iliac spine

Sartorius

Pectineus

Gracilis

Patella

(a)

Pectineus
(cut)

Femur

O = origin
I = insertion

(b)

M. lu..

■ **Figure 6.26 (a) Muscles of the anterior thigh, superficial group
and (b) exposure of adductors**

Muscles That Act on the Thigh (Femur)

Adductor brevis

Origin: _____

Insertion: _____

Action: _____

Adductor longus

Origin: _____

Insertion: _____

Action: _____

Adductor magnus

Origin: _____

Insertion: _____

Action: _____

Tensor fasciae latae

Origin: _____

Insertion: _____

Action: _____

IDENTIFY

Gracilis
Rectus femoris
Sartorius
Vastus lateralis
Vastus medialis

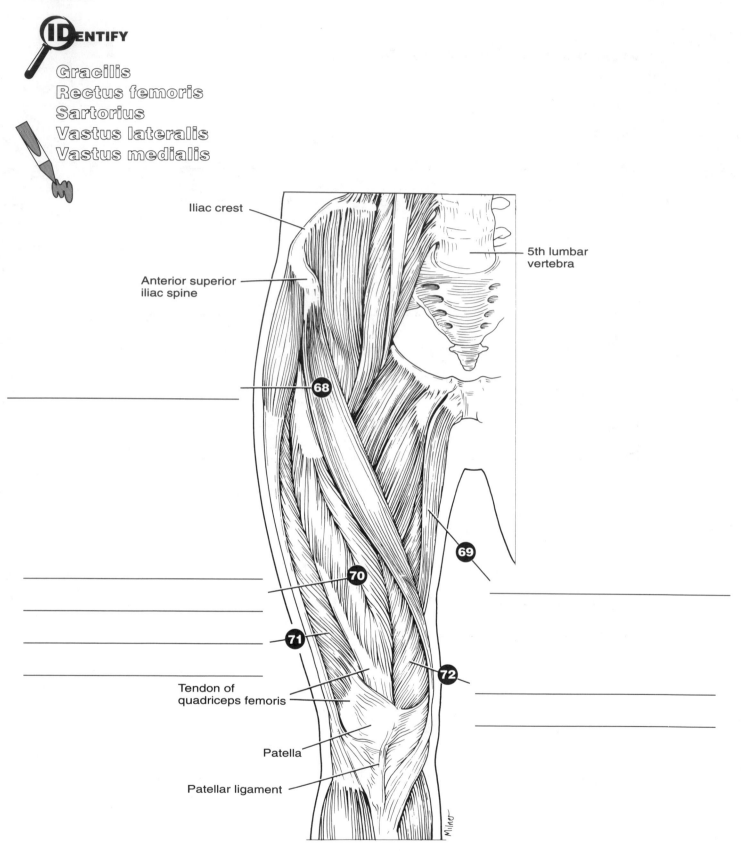

Iliac crest

Anterior superior
iliac spine

5th lumbar
vertebra

68

69

70

71

72

Tendon of
quadriceps femoris

Patella

Patellar ligament

■ **Figure 6.27 Muscles of the anterior thigh, superficial group**

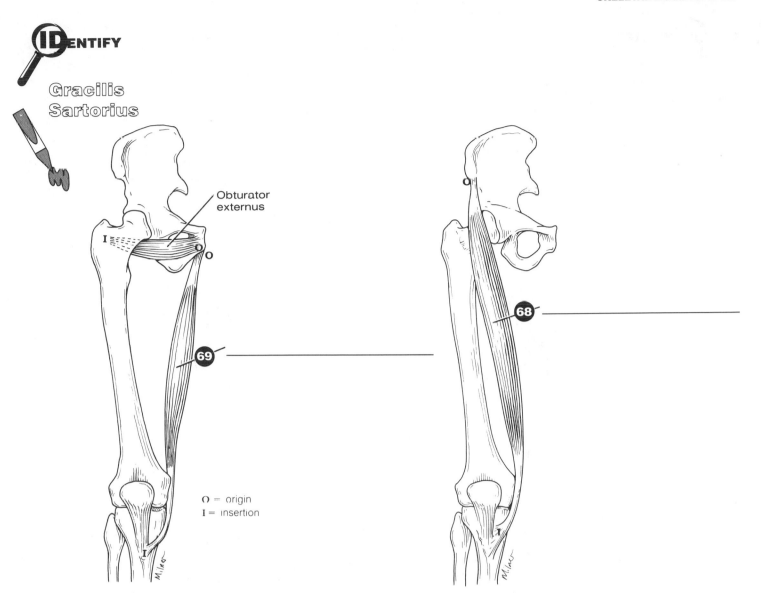

IDENTIFY

Gracilis
Sartorius

■ **Figure 6.28 Exposure of the gracilis and sartorius muscles**

Muscles That Act on the Leg (Muscles of the Thigh)

Gracilis

Origin:—————————————————————

Insertion:—————————————————————

Action:—————————————————————

Sartorius

Origin:—————————————————————

Insertion:—————————————————————

Action:—————————————————————

IDENTIFY

Rectus femoris
Vastus lateralis

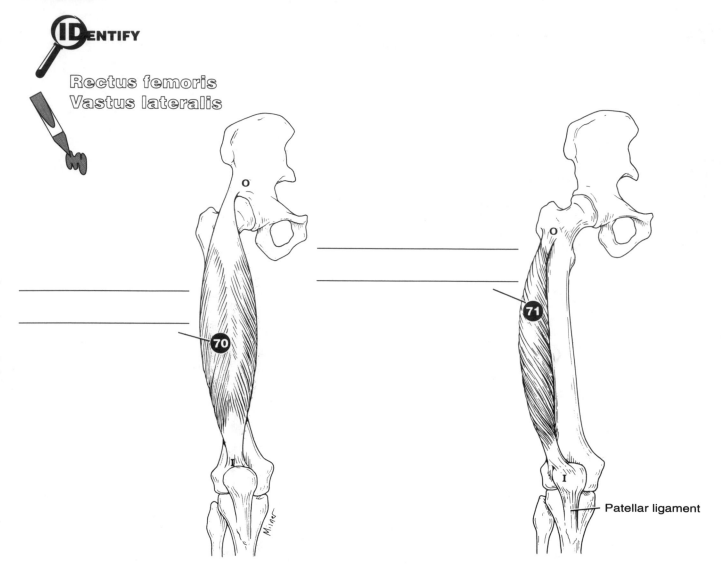

■ **Figure 6.29 Components of the quadriceps femoris**

Muscles That Act on the Leg (Muscles of the Thigh)

Rectus femoris

Origin: _____

Insertion: _____

Action: _____

Vastus lateralis

Origin: _____

Insertion: _____

Action: _____

IDENTIFY

Vastus intermedius
Vastus medialis

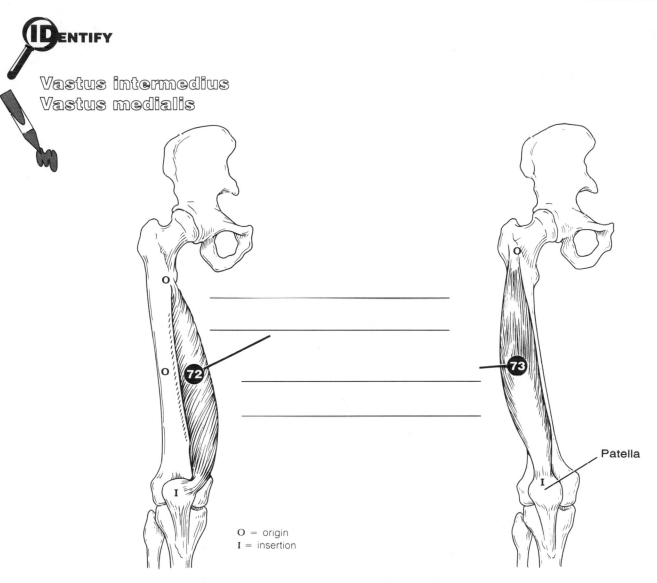

O = origin
I = insertion

Patella

■ **Figure 6.30 Components of the quadriceps femoris**

Muscles That Act on the Leg (Muscles of the Thigh)

Vastus intermedius

Origin: _____

Insertion: _____

Action: _____

Vastus medialis

Origin: _____

Insertion: _____

Action: _____

Q214-219

IDENTIFY

Biceps femoris
Semimembranosus
Semitendinosus

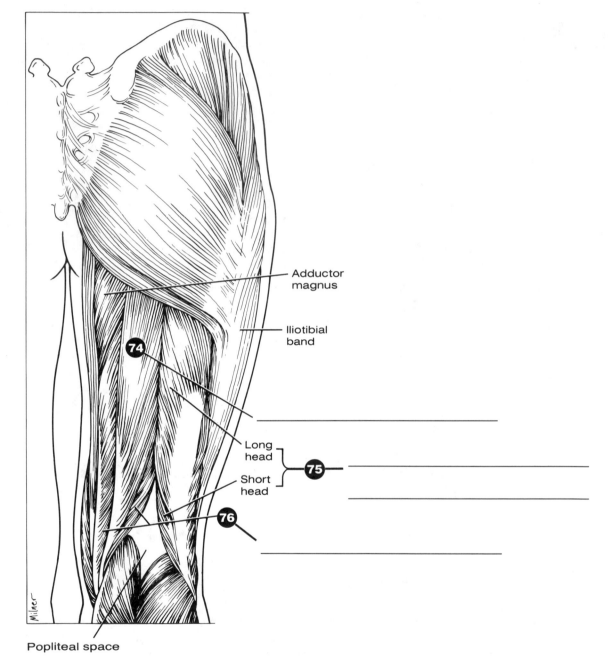

Adductor
magnus

Iliotibial
band

74

Long
head

Short
head

75

76

Popliteal space

Milner

■ **Figure 6.31 Muscles of the posterior thigh, superficial group**

Muscles That Act on the Leg (Muscles of the Thigh)

Biceps femoris

Origin: _____

Insertion: _____

Action: _____

Semimembranosus

Origin: _____

Insertion: _____

Action: _____

Semitendinosus

Origin: _____

Insertion: _____

Action: _____

IDENTIFY

Extensor digitorum longus
Tibialis anterior

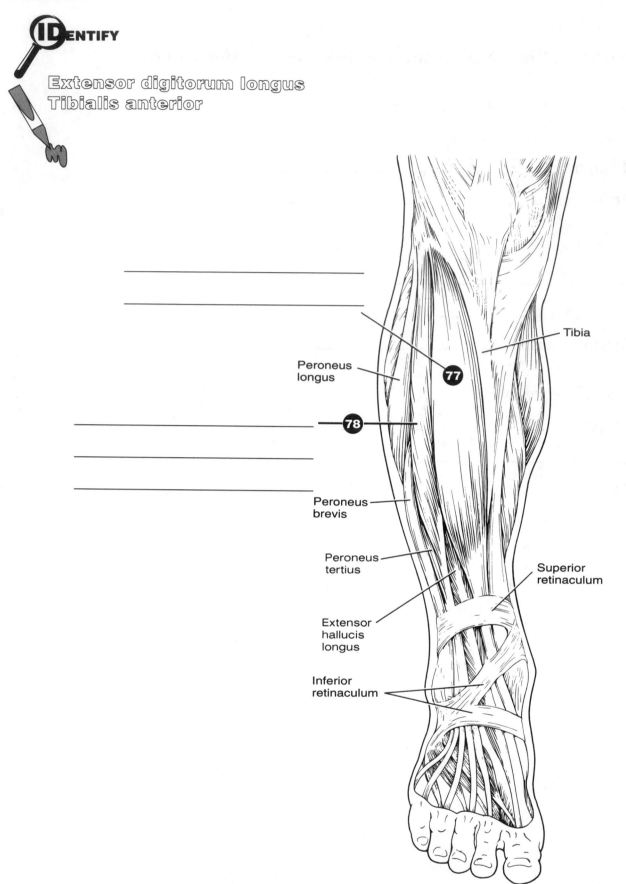

Tibia

Peroneus
longus

77

78

Peroneus
brevis

Peroneus
tertius

Superior
retinaculum

Extensor
hallucis
longus

Inferior
retinaculum

■ **Figure 6.32 Muscles of the anterior leg, superficial group**

IDENTIFY

Extensor digitorum longus
Tibialis anterior

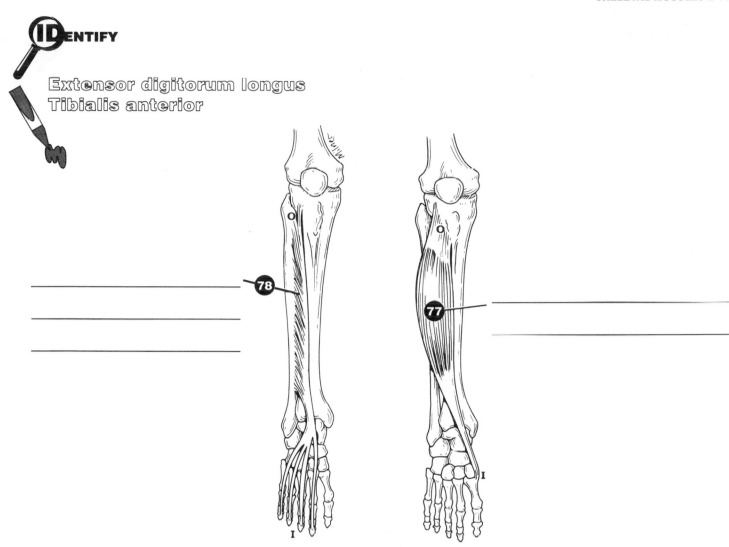

■ **Figure 6.33 Exposure of the tibialis anterior and extensor digitorum longus muscles**

Muscles That Act on the Foot and Toes (Muscles of the Leg)

Extensor digitorum longus

Origin:_____

Insertion:_____

Action:_____

Tibialis anterior

Origin:_____

Insertion:_____

Action:_____

Q229-234

IDENTIFY

Peroneus brevis
Peroneus longus

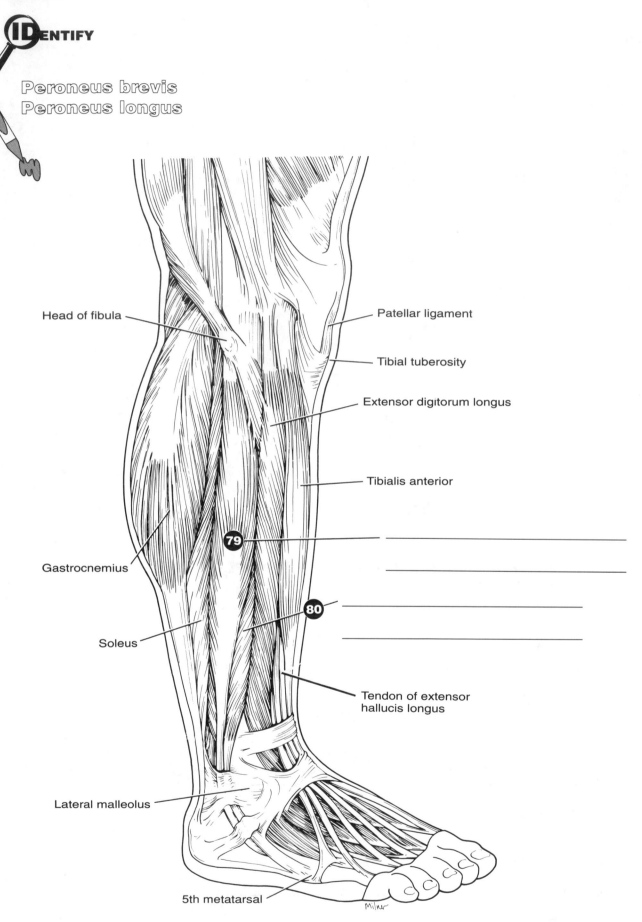

Head of fibula

Patellar ligament

Tibial tuberosity

Extensor digitorum longus

Tibialis anterior

79 _____

_____

Gastrocnemius

80 _____

_____

Soleus

Tendon of extensor
hallucis longus

Lateral malleolus

5th metatarsal

Milner

■ **Figure 6.34 Muscles of the lateral leg, superficial group**

IDENTIFY

Peroneus brevis
Peroneus longus

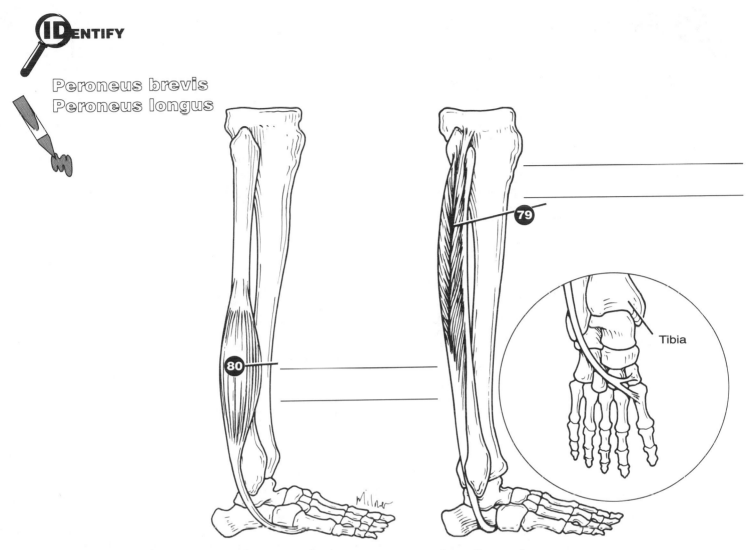

Tibia

■ **Figure 6.35 Exposure of the peroneus brevis and peroneus longus muscles**

Muscles That Act on the Foot and Toes (Muscles of the Leg)

Peroneus brevis

Origin: _____

Insertion: _____

Action: _____

Peroneus longus

Origin: _____

Insertion: _____

Action: _____

Q235-240

IDENTIFY

Extensor hallucis longus
Peroneus tertius

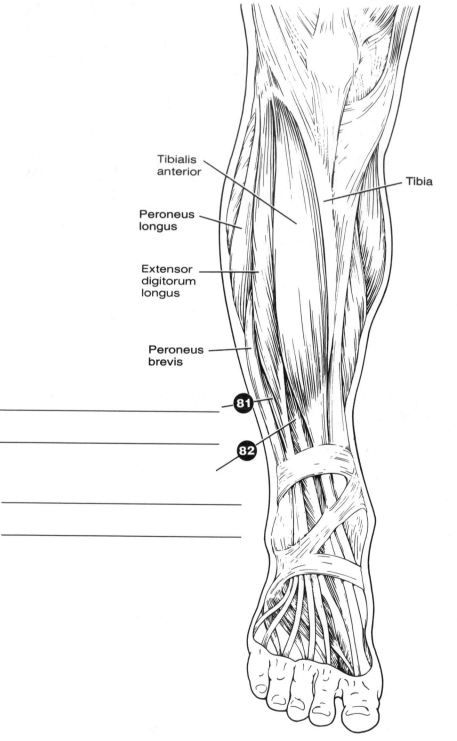

Tibialis
anterior

Tibia

Peroneus
longus

Extensor
digitorum
longus

Peroneus
brevis

81

82

■ **Figure 6.36 Muscles of the anterior leg, deep group**

IDENTIFY

Extensor hallucis longus
Peroneus tertius

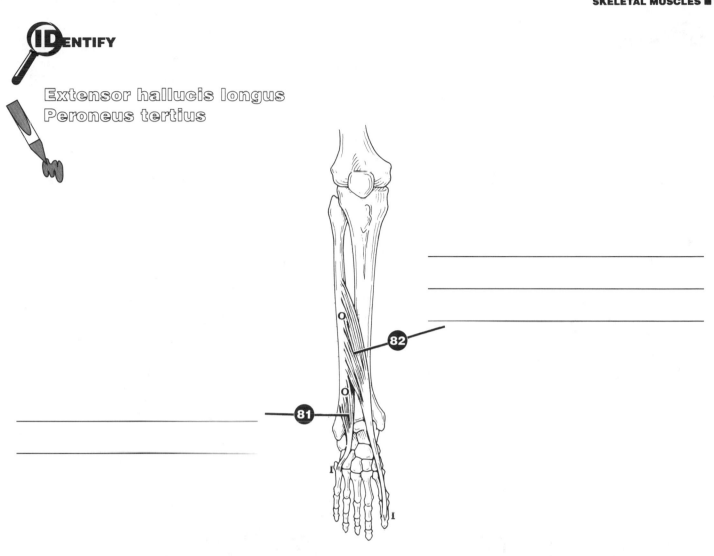

■ **Figure 6.37 Exposure of the peroneus tertius and extensor hallucis longus muscles**

Muscles That Act on the Foot and Toes (Muscles of the Leg)

Extensor hallucis longus

Origin: _____

Insertion: _____

Action: _____

Peroneus tertius

Origin: _____

Insertion: _____

Action: _____

Q241-
246

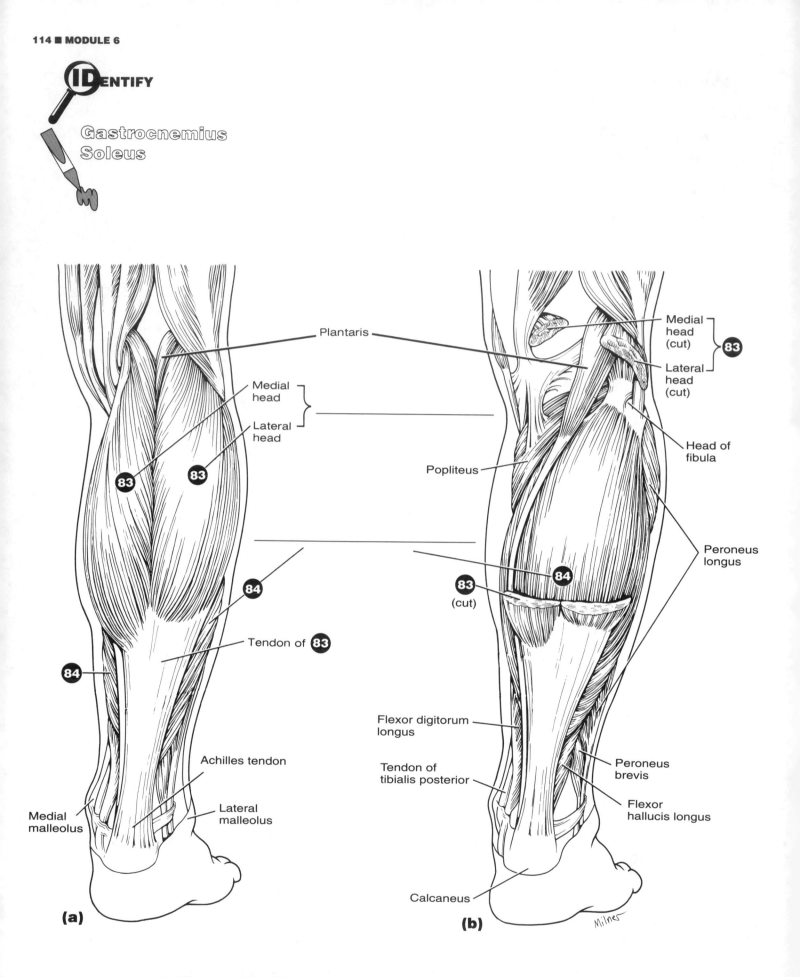

IDENTIFY

Gastrocnemius
Soleus

■ **Figure 6.38 Muscles of the posterior leg: (a) superficial
and (b) middle groups**

Muscles That Act on the Foot and Toes
(Muscles of the Leg)

Gastrocnemius

Origin: _____

Insertion: _____

Action: _____

Soleus

Origin: _____

Insertion: _____

Action: _____

IDENTIFY

Flexor hallucis longus
Tibialis posterior

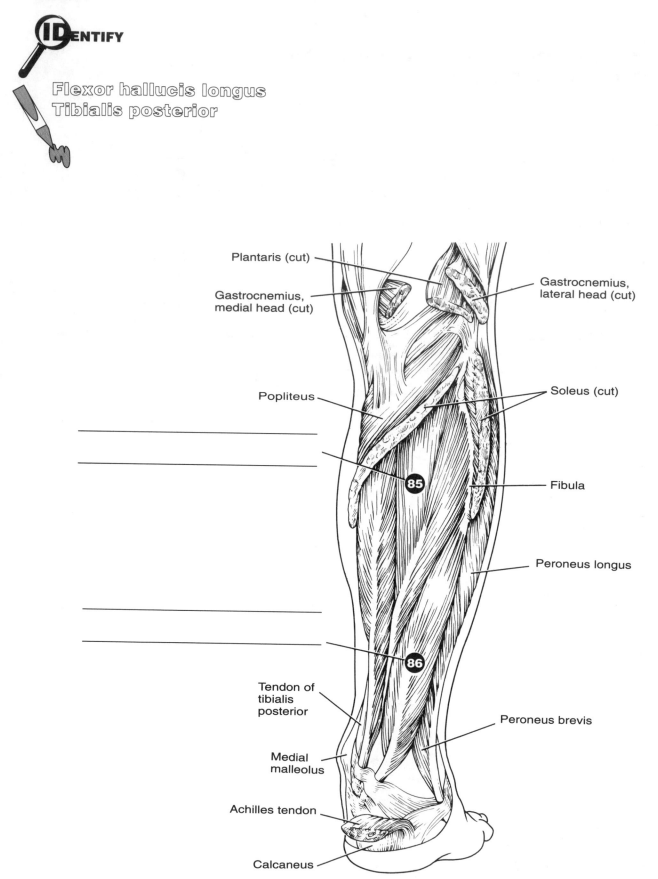

Plantaris (cut)

Gastrocnemius,
medial head (cut)

Gastrocnemius,
lateral head (cut)

Popliteus

Soleus (cut)

85

Fibula

Peroneus longus

86

Tendon of
tibialis
posterior

Peroneus brevis

Medial
malleolus

Achilles tendon

Calcaneus

■ **Figure 6.39 Muscles of the posterior leg, deep group**

IDENTIFY

Flexor hallucis longus
Tibialis posterior

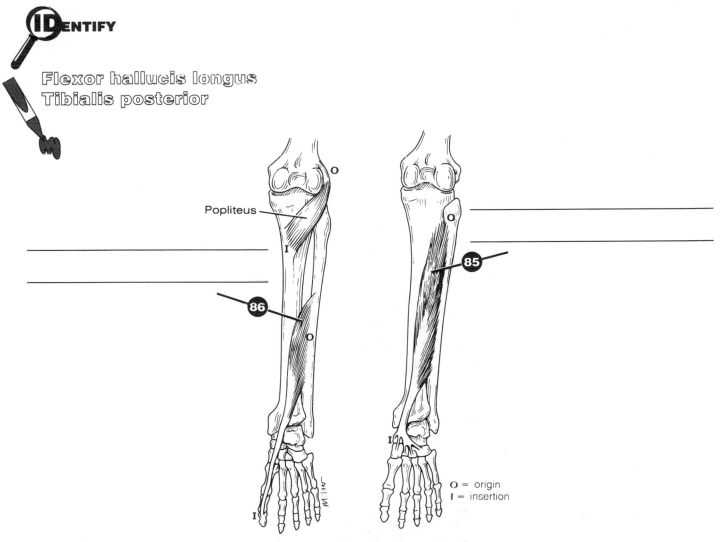

■ **Figure 6.40 Exposure of the flexor hallucis longus and tibialis posterior muscles**

Muscles That Act on the Foot and Toes (Muscles of the Leg)

Flexor hallucis longus

Origin: _____

Insertion: _____

Action: _____

Tibialis posterior

Origin: _____

Insertion: _____

Action: _____

IDENTIFY

Flexor digitorum longus

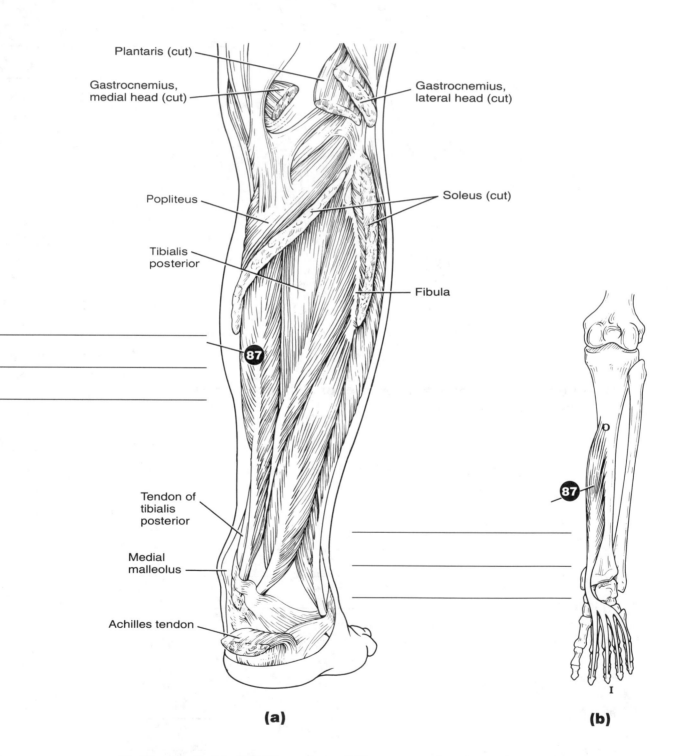

Plantaris (cut)

Gastrocnemius, medial head (cut)

Gastrocnemius, lateral head (cut)

Popliteus

Soleus (cut)

Tibialis posterior

Fibula

87

Tendon of tibialis posterior

Medial malleolus

87

Achilles tendon

(a)

(b)

■ **Figure 6.41 (a) Muscles of the posterior leg, deep group and (b) exposure of the flexor digitorum longus**

IDENTIFY

Flexor digitorum brevis

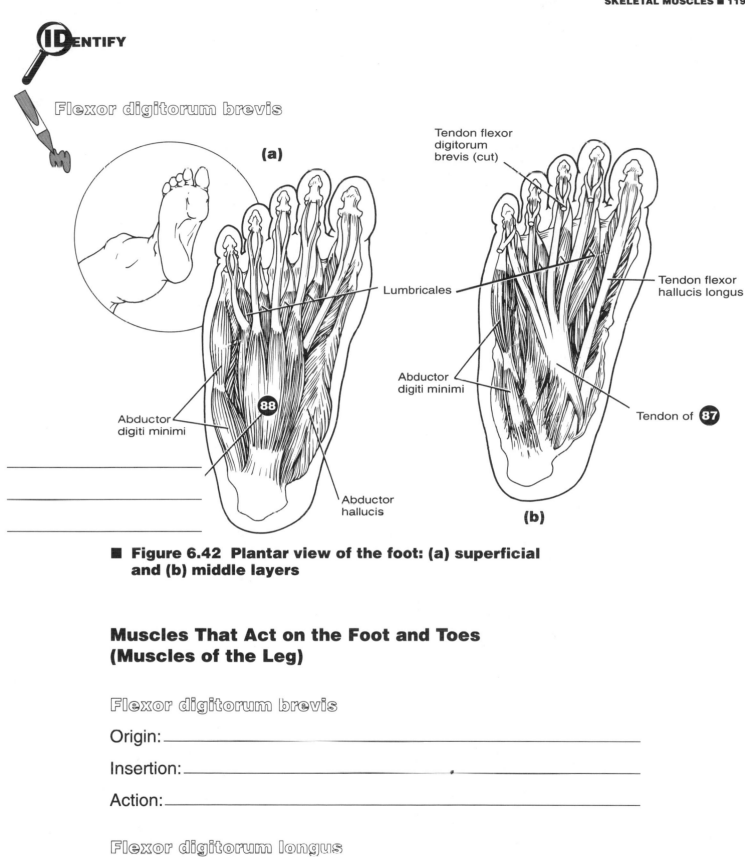

■ **Figure 6.42 Plantar view of the foot: (a) superficial and (b) middle layers**

Muscles That Act on the Foot and Toes (Muscles of the Leg)

Flexor digitorum brevis

Origin: _____

Insertion: _____

Action: _____

Flexor digitorum longus

Origin: _____

Insertion: _____

Action: _____

Q259-264

MODULE 7

The Circulatory System

IDENTIFY

**All heart valves
(Hint: Bicuspid, etc.)
Chordae tendineae
Coronary sinus opening
Interventricular septum
Papillary muscle
Parietal pericardium
Trabeculae carnae
Visceral pericardium**

Aortic arch
Atria: left, right
Descending aorta
Endocardium
Inferior vena cava
Myocardium
Pulmonary trunk
Superior vena cava
Ventricles: left, right

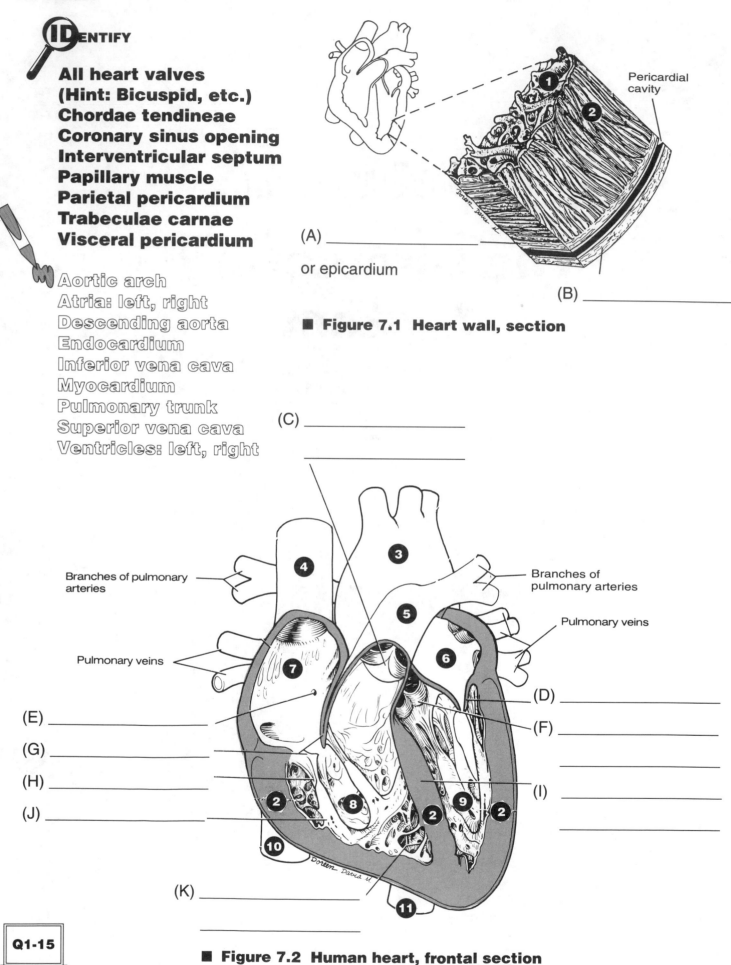

Pericardial cavity

(A) _____

or epicardium

(B) _____

■ **Figure 7.1 Heart wall, section**

(C) _____

_____

Branches of pulmonary arteries

Branches of pulmonary arteries

Pulmonary veins

Pulmonary veins

(E) _____

(G) _____

(H) _____

(J) _____

(D) _____

(F) _____

_____

(I) _____

_____

(K) _____

_____

■ **Figure 7.2 Human heart, frontal section**

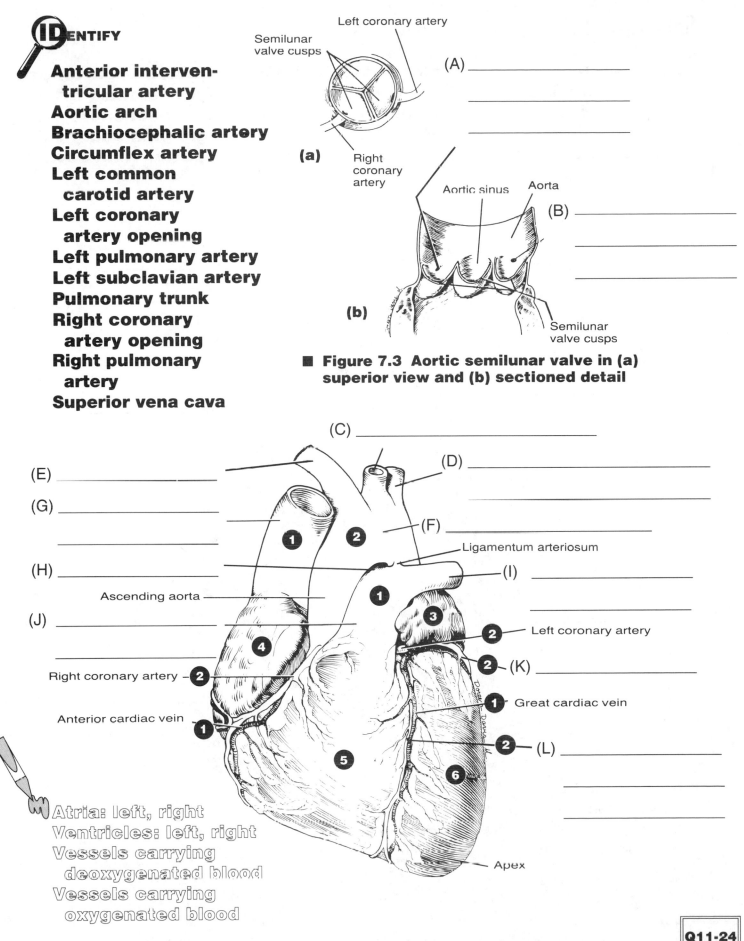

ID**ENTIFY**

Anterior interven-
tricular artery
Aortic arch
Brachiocephalic artery
Circumflex artery
Left common
carotid artery
Left coronary
artery opening
Left pulmonary artery
Left subclavian artery
Pulmonary trunk
Right coronary
artery opening
Right pulmonary
artery
Superior vena cava

Left coronary artery

Semilunar
valve cusps

(A) _____

_____

_____

(a)

Right
coronary
artery

Aortic sinus   Aorta

(B) _____

_____

_____

(b)

Semilunar
valve cusps

■ **Figure 7.3 Aortic semilunar valve in (a)**
superior view and (b) sectioned detail

(C) _____

(D) _____

(E) _____

(G) _____

_____

(H) _____

(F) _____

Ligamentum arteriosum

Ascending aorta

(I) _____

_____

(J) _____

_____

Right coronary artery — ②

Anterior cardiac vein

Left coronary artery

② (K) _____

① Great cardiac vein

② (L) _____

_____

_____

Apex

Atria: left, right
Ventricles: left, right
Vessels carrying
    deoxygenated blood
Vessels carrying
    oxygenated blood

■ **Figure 7.4 Vessels of the heart, anterior view**

IDENTIFY

Aortic arch
Brachiocephalic artery
Circumflex artery
Coronary sinus
Inferior vena cava

Left common carotid
  artery
Left pulmonary artery
Left pulmonary veins
Left subclavian artery

Ligamentum arteriosum
Right coronary artery
Right pulmonary artery
Right pulmonary veins
Superior vena cava

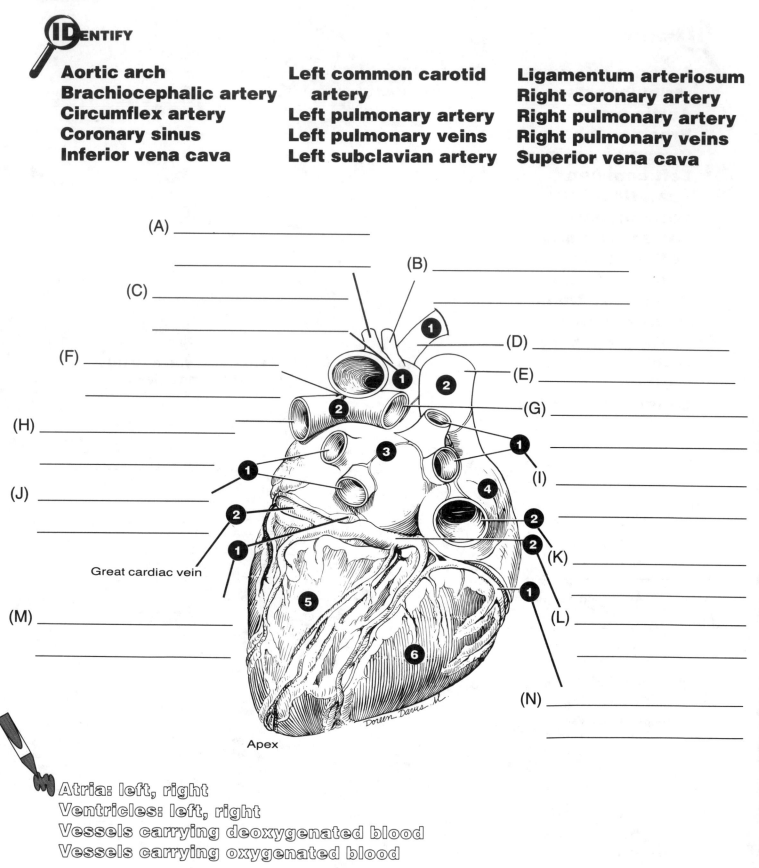

(A) _____

_____

(C) _____

_____

(F) _____

_____

(H) _____

_____

(J) _____

_____

Great cardiac vein

(M) _____

_____

Apex

(B) _____

_____

(D) _____

(E) _____

(G) _____

_____

(I) _____

(K) _____

_____

(L) _____

_____

(N) _____

_____

Atria: left, right
Ventricles: left, right
Vessels carrying deoxygenated blood
Vessels carrying oxygenated blood

Q12-24

■ **Figure 7.5 Vessels of the heart, posterior view**

Doreen Davis M.

ID ENTIFY

Ductus arteriosus
External carotid
  artery
Facial artery
Foramen ovalis
Internal carotid
  artery
Right common
  carotid artery
Right subclavian
  artery
Superficial temporal
  artery
Vertebral artery

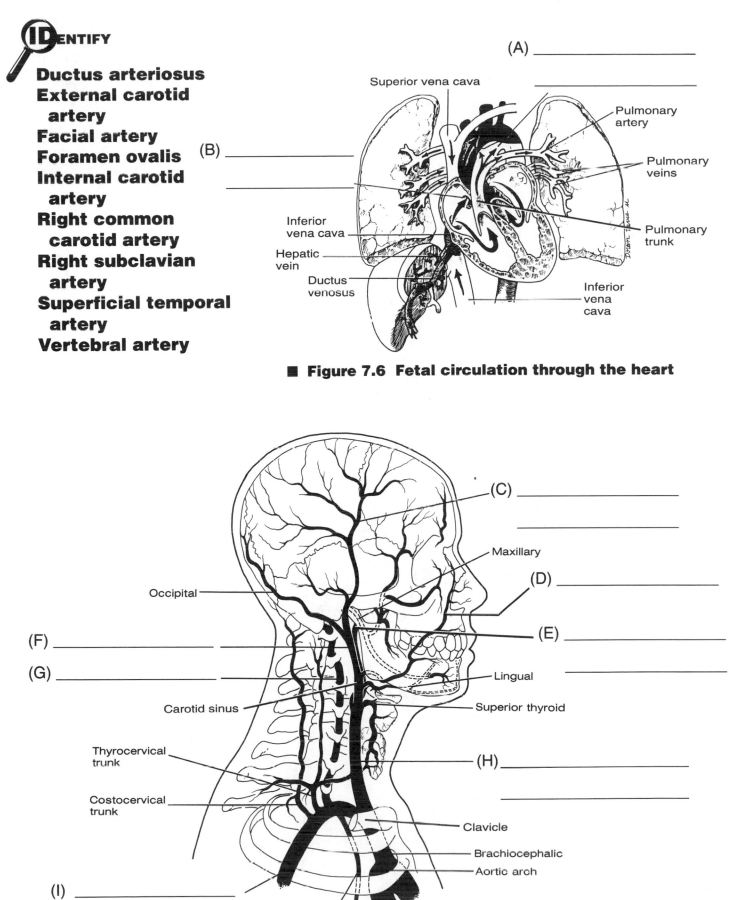

(A) _____
_____

Superior vena cava

Pulmonary artery

(B) _____

Pulmonary veins

Inferior vena cava

Hepatic vein

Ductus venosus

Pulmonary trunk

Inferior vena cava

■ Figure 7.6 Fetal circulation through the heart

(C) _____
_____

Maxillary

(D) _____

Occipital

(E) _____

(F) _____

(G) _____

Lingual

Carotid sinus

Superior thyroid

Thyrocervical trunk

(H) _____
_____

Costocervical trunk

Clavicle

Brachiocephalic

Aortic arch

(I) _____
_____

Internal thoracic

Q22-30

■ Figure 7.7 Branches of the brachiocephalic artery

IDENTIFY

Basilar artery
Internal carotid artery
Vertebral artery

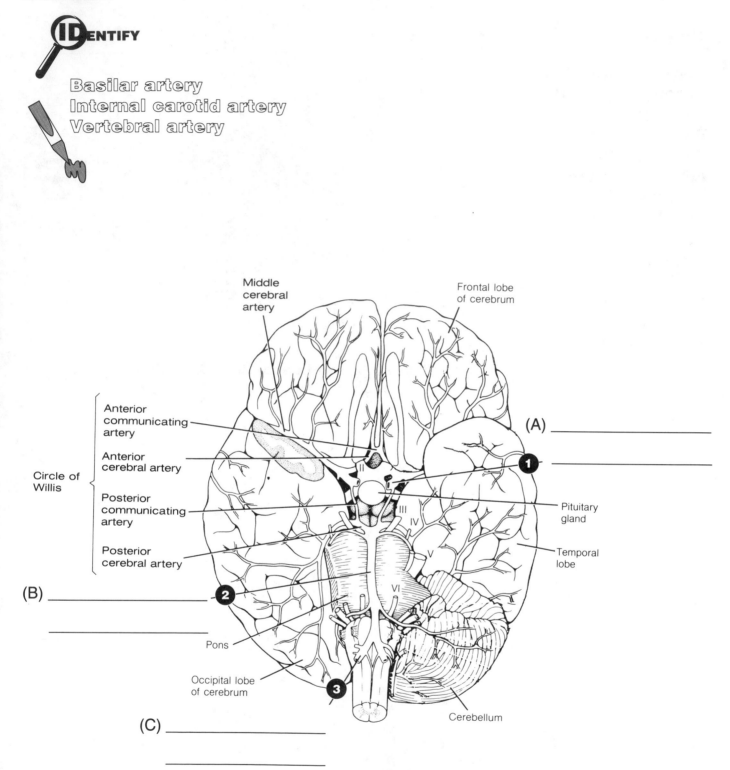

■ Figure 7.8 Arteries serving the brain

IDENTIFY

Axillary artery
Brachial artery
Left common carotid artery
Radial artery
Subclavian artery
Ulnar artery

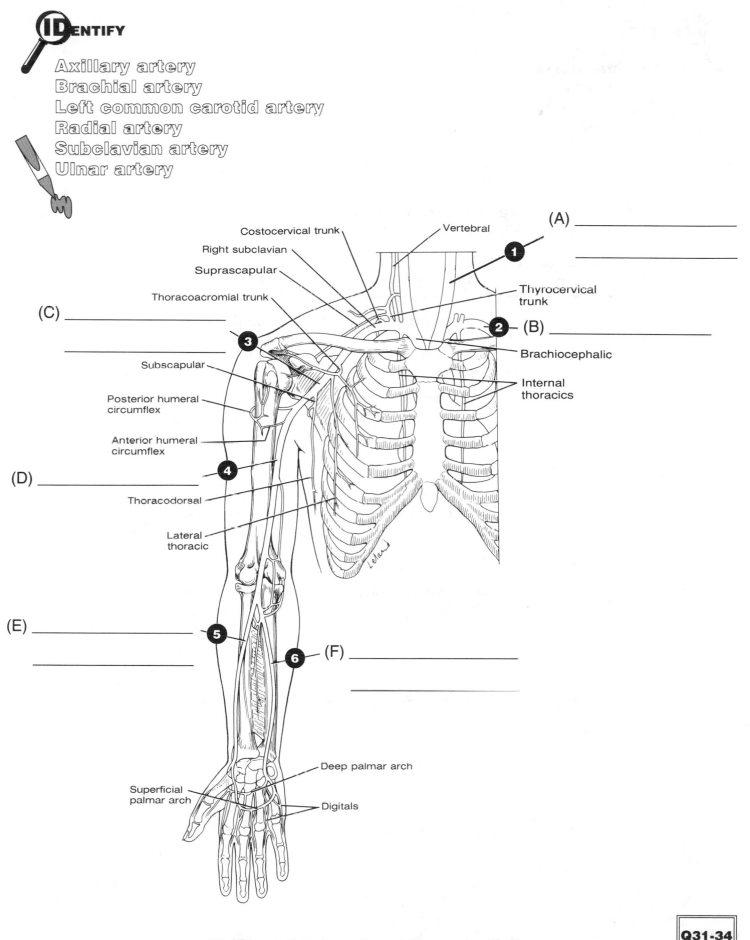

Costocervical trunk
Vertebral
Right subclavian
Suprascapular
Thoracoacromial trunk
Thyrocervical trunk
Brachiocephalic
Internal thoracics
Subscapular
Posterior humeral circumflex
Anterior humeral circumflex
Thoracodorsal
Lateral thoracic
Deep palmar arch
Superficial palmar arch
Digitals

(A) _____
(B) _____
(C) _____
(D) _____
(E) _____
(F) _____

■ Figure 7.9 Arteries of the upper limb

Q31-34

ENTIFY

Abdominal aorta
Arch of aorta
Ascending aorta
Thoracic aorta

■ **Table 7.1 Anatomical Areas of the Aorta**
  (Refer to Figures 7.3, 7.4, 7.5, 7.10 and 7.11)

Aortic segment	Description	Organ(s) served
(A)	Between heart and arch	Heart only, via coronary arteries
(B)	Between ascending and descending aortas	Head, neck, and arms via brachiocephalic, subclavians, and carotids
(C)	Descending aorta between arch and diaphragm	Esophagus, trachea and bronchi, ribs and intercostal muscles, diaphragm
(D)	Descending aorta below the diaphragm	Liver, spleen, pancreas, small and large intestines, kidneys, gonads, lumbar musculature, bladder, genitalia, lower appendages

Q35-36

IDENTIFY

Anterior intercostal arteries
Brachiocephalic artery
Common carotid arteries
Left subclavian artery
Posterior intercostal arteries
Right subclavian artery
Thoracic aorta

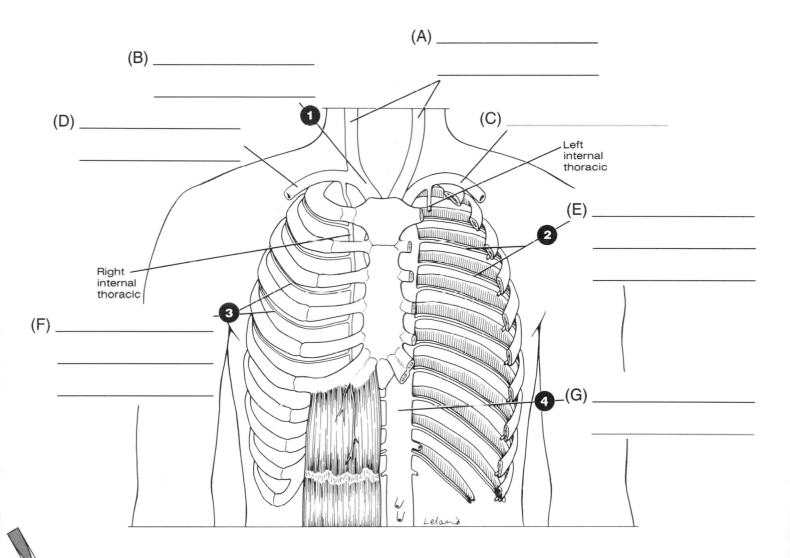

(A) _____

(B) _____

(D) _____

(C) _____

Left
internal
thoracic

(E) _____

Right
internal
thoracic

(F) _____

(G) _____

Leland

Anterior intercostal arteries
Brachiocephalic artery
Posterior intercostal arteries
Thoracic aorta

■ **Figure 7.10 Arteries of the thoracic region**

Q37

IDENTIFY

Abdominal aorta
Celiac artery
Common iliac artery
Inferior mesenteric artery

Renal artery
Spermatic/ovarian artery
Superior mesenteric artery

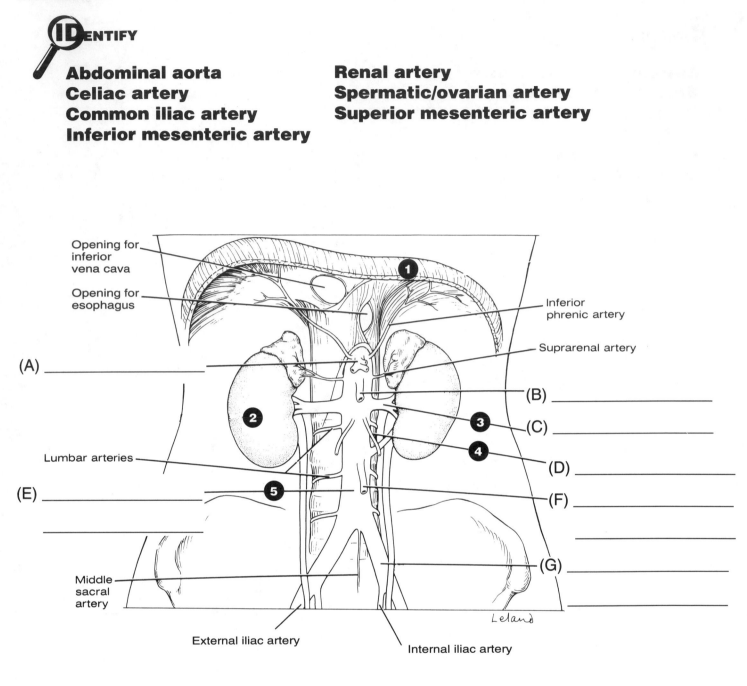

Opening for inferior vena cava

Opening for esophagus

Inferior phrenic artery

Suprarenal artery

(A) _____

(B) _____

(C) _____

(D) _____

Lumbar arteries

(E) _____

(F) _____

_____

(G) _____

Middle sacral artery

External iliac artery

Internal iliac artery

Leland

Abdominal aorta
Diaphragm
Kidney
Renal artery
Spermatic/ovarian artery

■ Figure 7.11 Branches of the abdominal aorta

Q38-47

IDENTIFY

Hepatic artery
Left gastric artery
Splenic artery

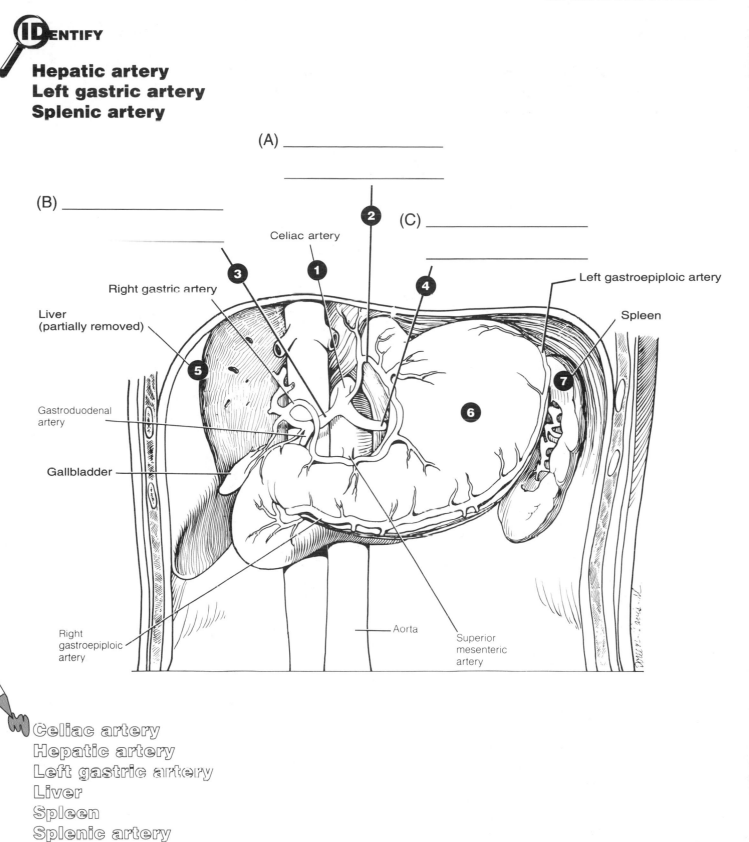

(A) _____

_____

(B) _____

_____

Celiac artery

(C) _____

_____

2

3

1

4

Right gastric artery

Left gastroepiploic artery

Liver
(partially removed)

5

7

Spleen

Gastroduodenal
artery

6

Gallbladder

Right
gastroepiploic
artery

Aorta

Superior
mesenteric
artery

Celiac artery
Hepatic artery
Left gastric artery
Liver
Spleen
Splenic artery
Stomach

■ Figure 7.12 Branches of the celiac artery

Q40-42

IDENTIFY

Common iliac artery   **Internal iliac artery**
External iliac artery   **Superior mesenteric artery**
Inferior mesenteric artery

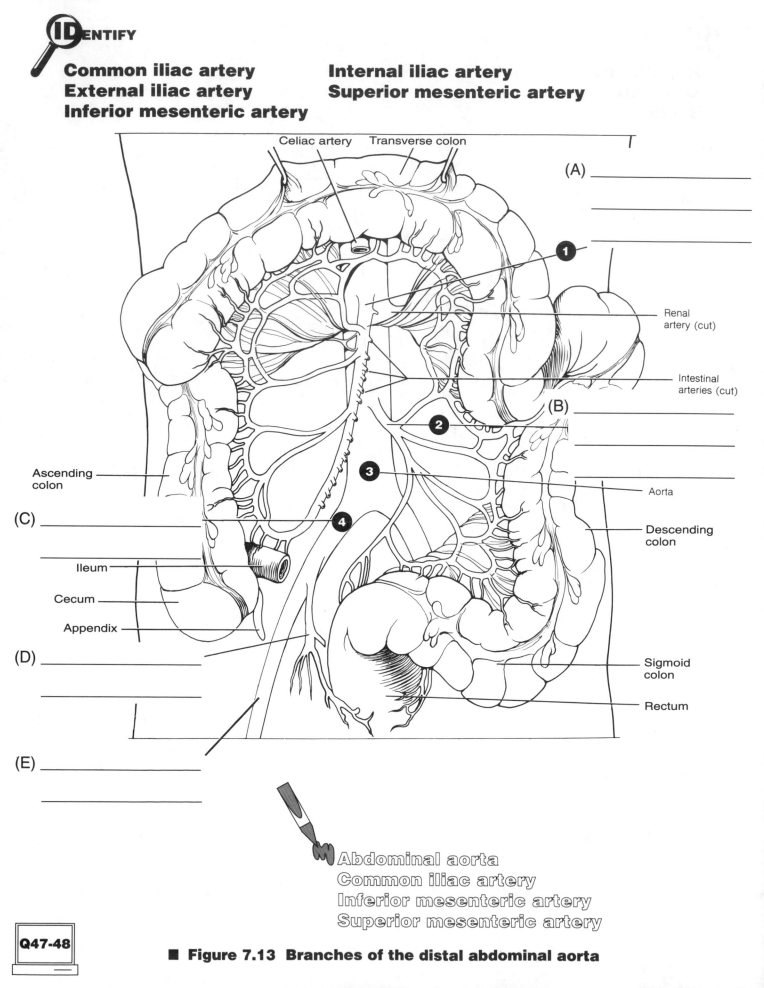

Celiac artery   Transverse colon

(A) _____

_____

1 _____

Renal artery (cut)

Intestinal arteries (cut)

(B) _____

2 _____

3 _____

Aorta

Ascending colon

Descending colon

(C) _____

Ileum

4

Cecum

Appendix

Sigmoid colon

(D) _____

Rectum

(E) _____

_____

Abdominal aorta
Common iliac artery
Inferior mesenteric artery
Superior mesenteric artery

Q47-48

■ **Figure 7.13 Branches of the distal abdominal aorta**

IDENTIFY

Common iliac artery	Renal artery
Hepatic artery	Spermatic artery
Inferior mesenteric artery	Splenic artery
Inferior phrenic artery	Superior mesenteric artery
Left gastric artery	Suprarenal artery
Ovarian artery	

■ **Table 7.2 Branches of the Abdominal Aorta Summarized**
   (Refer to Figures 7.11, 7.12 and 7.13)

Abdominal Vessel	Descriptive Comments (Organs Supplied, etc.)
(A)	Diaphragm
Celiac trunk	Divides immediately, forming vessels (B), (C), and (D)
(B)	Lesser curvature of stomach; lower esophagus
(C)	Liver and gallbladder; duodenum, stomach, pancreas
(D)	Spleen, stomach, pancreas
(E)	Small intestine, cecum, ascending and transverse colon
(F)	Adrenal glands
(G)	Kidneys
(H)	Ovaries
(I)	Testes
(J)	Transverse, descending, and sigmoid colon; rectum
(K)	Divides to send an internal iliac artery into the pelvic basin and an external iliac artery into the leg

Q38-48

IDENTIFY

Common iliac
External iliac
Femoral
Internal iliac
Popliteal

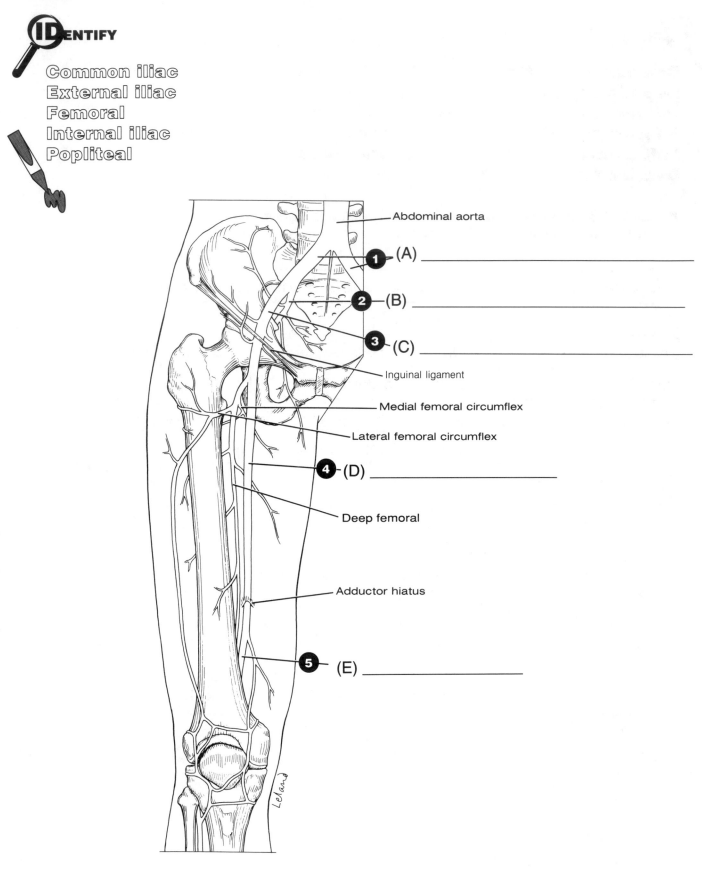

Abdominal aorta

1 —(A) _____

2 —(B) _____

3 —(C) _____

Inguinal ligament

Medial femoral circumflex

Lateral femoral circumflex

4 —(D) _____

Deep femoral

Adductor hiatus

5 (E) _____

■ **Figure 7.14 Arteries of the pelvis and thigh**

Q48-50

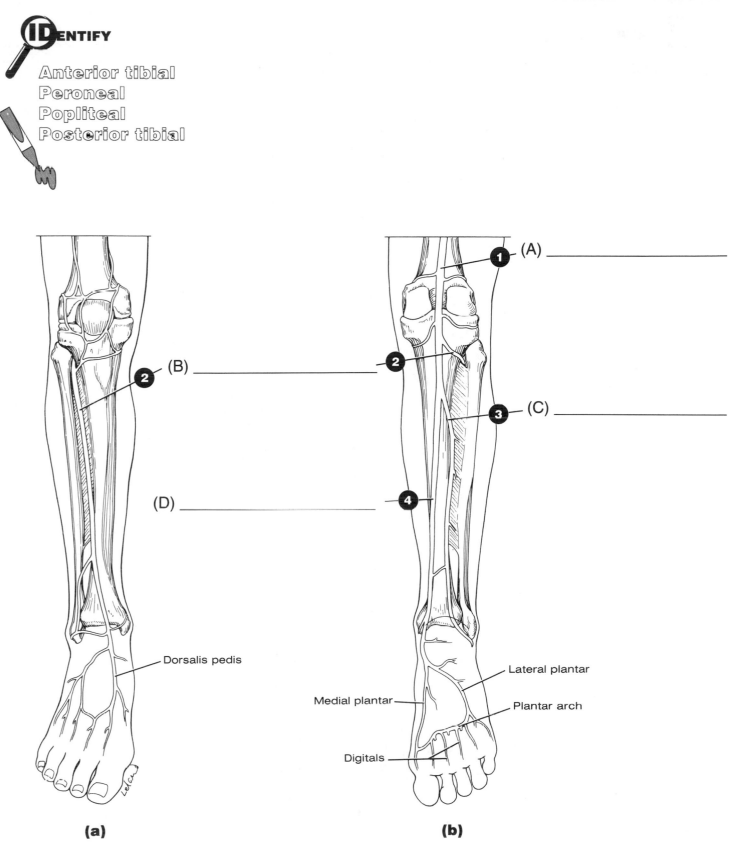

IDENTIFY

Anterior tibial
Peroneal
Popliteal
Posterior tibial

(A) _____

(B) _____

(C) _____

(D) _____

Dorsalis pedis

Lateral plantar

Medial plantar

Plantar arch

Digitals

(a)

(b)

■ **Figure 7.15 Arteries of the leg and foot in (a) anterior
and (b) posterior views**

Q51

IDENTIFY

Basilic
Brachiocephalic
Cephalic
External jugular
Greater saphenous

Hepatic
Inferior vena cava
Internal jugular
Median cubital
Superior vena cava

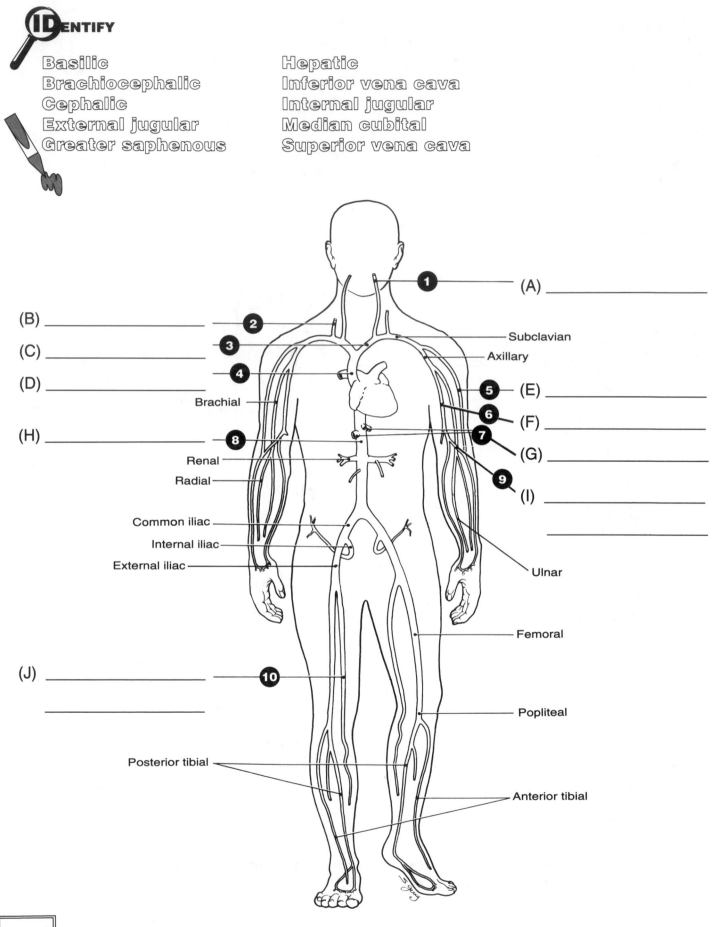

(A) _____

(B) _____

(C) _____

(D) _____

(H) _____

Subclavian

Axillary

(E) _____

(F) _____

(G) _____

(I) _____

_____

Brachial

Renal

Radial

Common iliac

Internal iliac

External iliac

Ulnar

Femoral

(J) _____

_____

Popliteal

Posterior tibial

Anterior tibial

■ Figure 7.16 Major systemic veins of the body

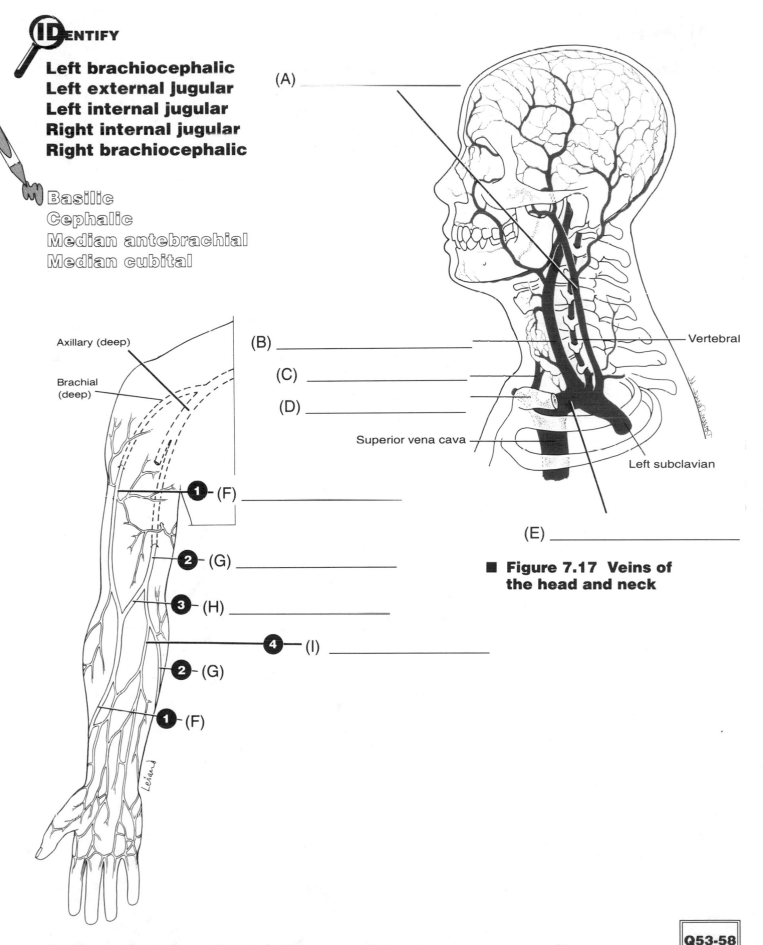

IDENTIFY

Left brachiocephalic
Left external Jugular
Left internal jugular
Right internal jugular
Right brachiocephalic

Basilic
Cephalic
Median antebrachial
Median cubital

(A) _____

Vertebral

(B) _____

(C) _____

(D) _____

Superior vena cava

Left subclavian

(E) _____

■ Figure 7.17 Veins of the head and neck

Axillary (deep)

Brachial (deep)

1 — (F) _____

2 — (G) _____

3 — (H) _____

4 — (I) _____

2 — (G)

1 — (F)

Lehew

■ Figure 7.18 Superficial veins, upper limb

Q53-58

IDENTIFY

Azygos
Hemiazygos
Left brachiocephalic
Left external jugular
Left internal jugular

Left subclavian
Posterior intercostals
Right axillary
Right brachiocephalic
Right subclavian

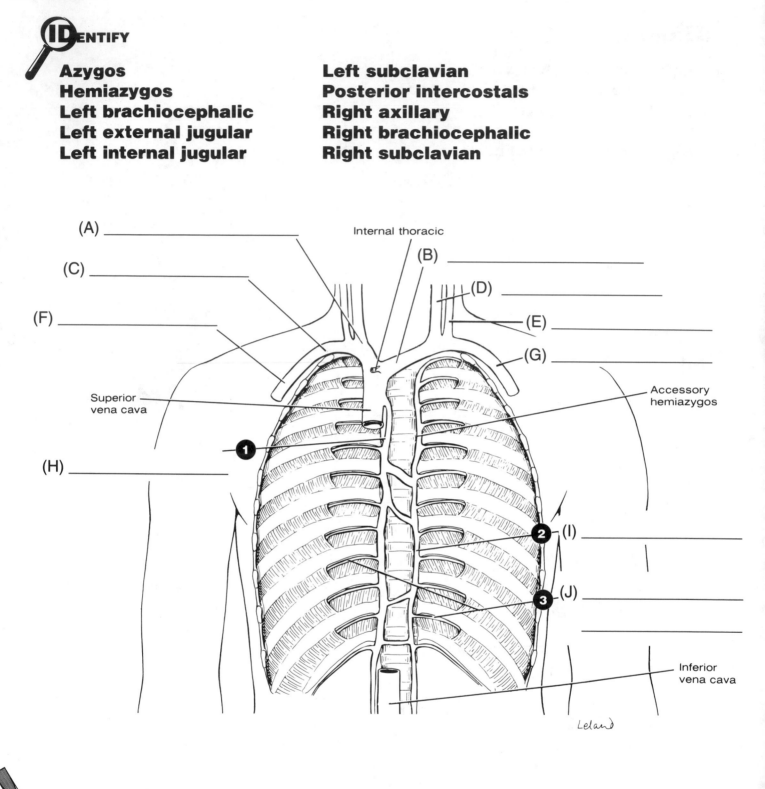

■ **Figure 7.19 Veins of the posterior thoracic wall**

Azygos
Hemiazygos
Posterior intercostals

Q55

IDENTIFY

Hepatic vein
Hepatic portal vein
Inferior mesenteric vein

Splenic vein
Superior mesenteric vein

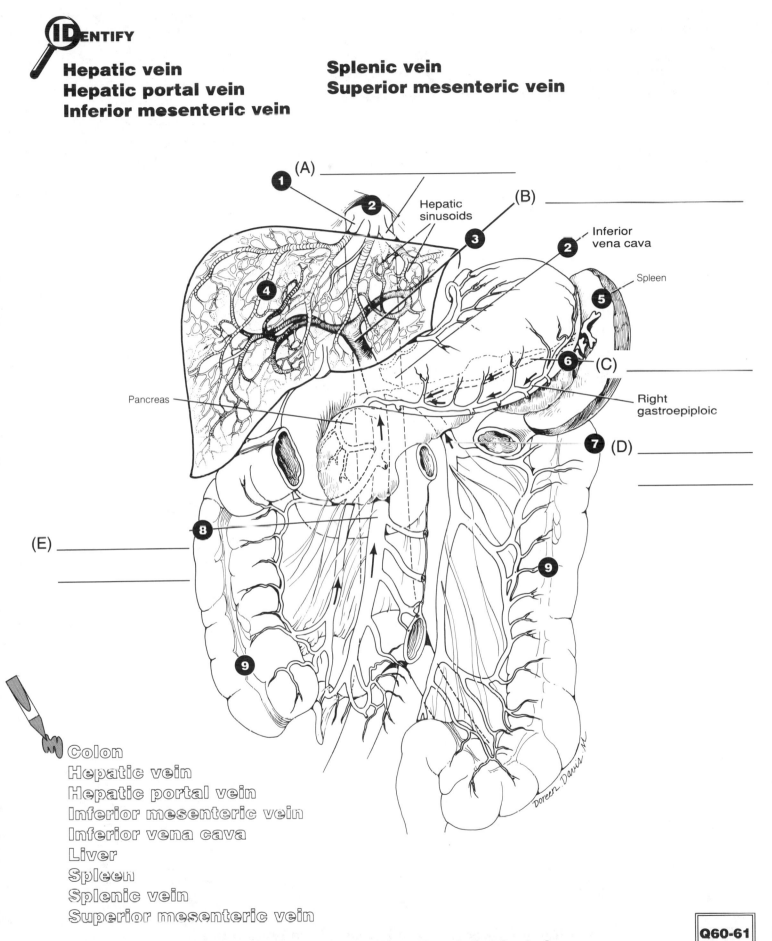

(A) _____

(B) _____

Hepatic sinusoids

Inferior vena cava

Spleen

Pancreas

Right gastroepiploic

(C) _____

(D) _____

(E) _____

Colon
Hepatic vein
Hepatic portal vein
Inferior mesenteric vein
Inferior vena cava
Liver
Spleen
Splenic vein
Superior mesenteric vein

Q60-61

■ **Figure 7.20 Veins draining the abdominal viscera**

IDENTIFY

Femoral (deep)   Lesser saphenous
Greater saphenous   Popliteal (deep)

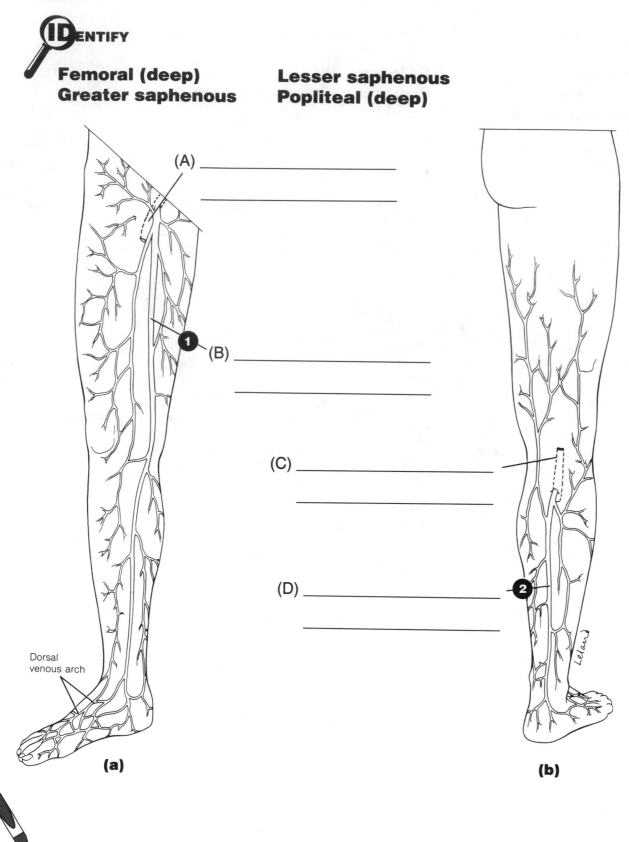

(A) _____

_____

1 (B) _____

_____

(C) _____

_____

(D) _____

_____

2

Dorsal
venous arch

(a)   (b)

Greater saphenous
Lesser saphenous

Q62

■ **Figure 7.21 Superficial veins of the lower limb in
(a) anterior and (b) posterior views**

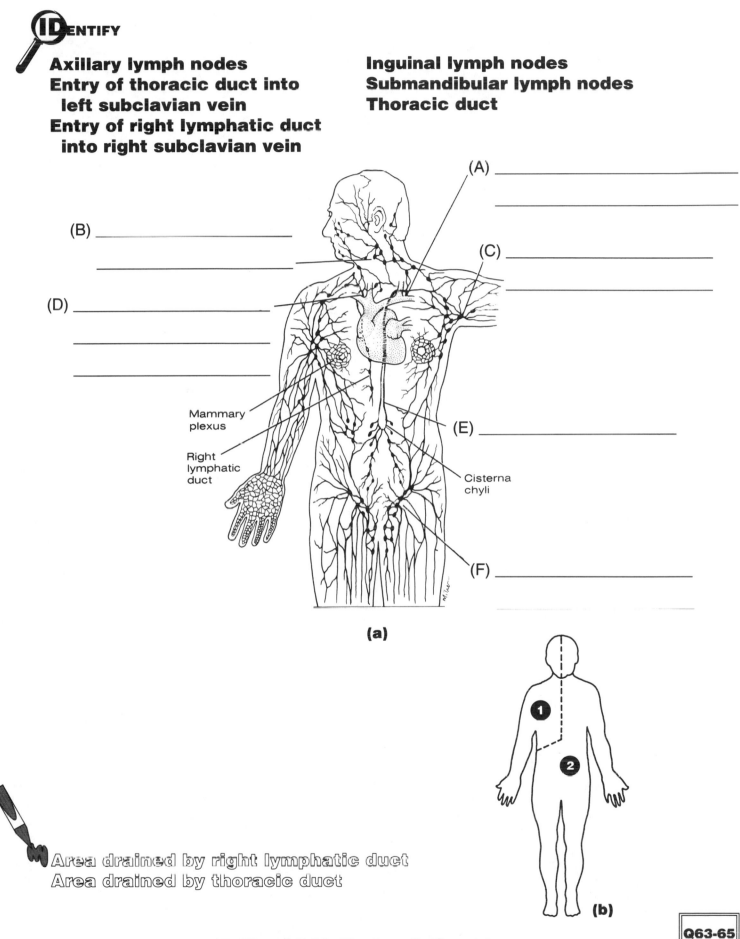

IDENTIFY

Axillary lymph nodes
Entry of thoracic duct into
  left subclavian vein
Entry of right lymphatic duct
  into right subclavian vein

Inguinal lymph nodes
Submandibular lymph nodes
Thoracic duct

(A) _____

_____

(B) _____

_____

(D) _____

_____

_____

(C) _____

_____

Mammary
plexus

Right
lymphatic
duct

(E) _____

Cisterna
chyli

(F) _____

(a)

Area drained by right lymphatic duct
Area drained by thoracic duct

1

2

(b)

Q63-65

■ Figure 7.22 The lymphatic system

MODULE

The Central Nervous System

IDENTIFY

Basal ganglia
Cerebellum
Cerebral cortex
Cerebral peduncles
Cerebrum
Corpora quadrigemina
Epithalamus

Hypothalamus
Medulla (oblongata)
Olfactory bulbs
Pons
Pineal (body)
Thalamus

■ **Table 8.1 Divisions and Subdivisions of the Brain**

Embryonic Divisions	Subdivisions	Adult Brain Structures
Forebrain (Prosencephalon)	Telencephalon	(A) _____
		Subparts:
		(B) _____
		(C) _____
		(D) _____
	Diencephalon	(E) _____
		(F) _____
		(G) _____
		Subpart:
		(H) _____
Midbrain (Mesencephalon)	Mesencephalon	(I) _____
		(J) _____
Hindbrain (Rhombencephalon)	Metencephalon	(K) _____
		(L) _____
		(M) _____

IDENTIFY

Central sulcus
Lateral sulcus
Longitudinal fissure
Parieto-occipital sulcus

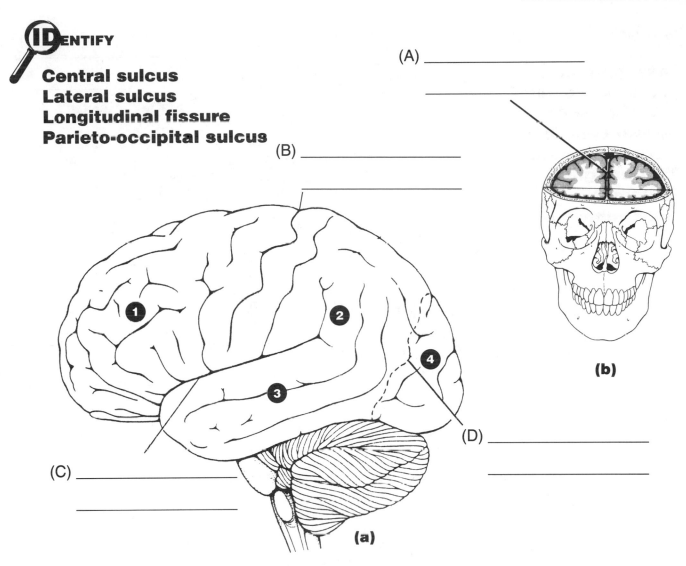

(A) _____

_____

(B) _____

_____

(C) _____

_____

(D) _____

_____

(b)

(a)

■ **Figure 8.1 (a) Lateral view and (b) frontal cut of the brain**

Lobes of the cerebrum	Functions
Frontal	(a) _____
	_____
Occipital	(b) _____
Parietal	(c) _____
	_____
	_____
Temporal	(d) _____
	_____

DEF

Q4-11

IDENTIFY

Basal ganglia
Corpus callosum
Hypothalamus
Lateral ventricles
Massa intermedia
Pituitary gland
Thalamus
Third ventricle

Corpus callosum
Pituitary gland

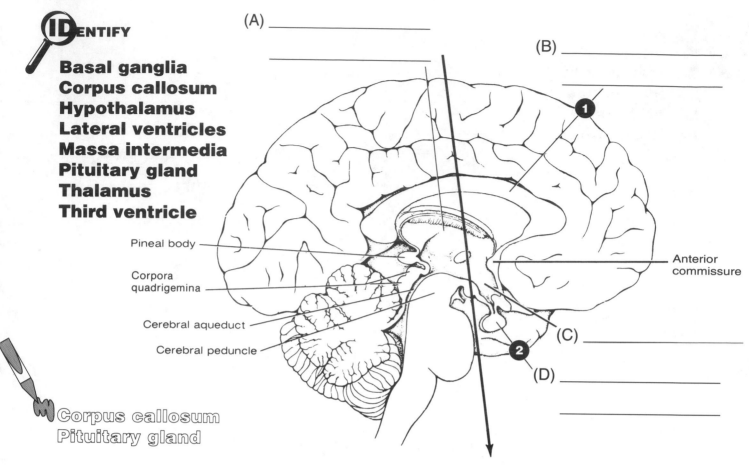

(A) _____

(B) _____

_____

Pineal body

Corpora quadrigemina

Cerebral aqueduct

Cerebral peduncle

Anterior commissure

(C) _____

(D) _____

_____

■ **Figure 8.2 Midsagittal section of the brain**

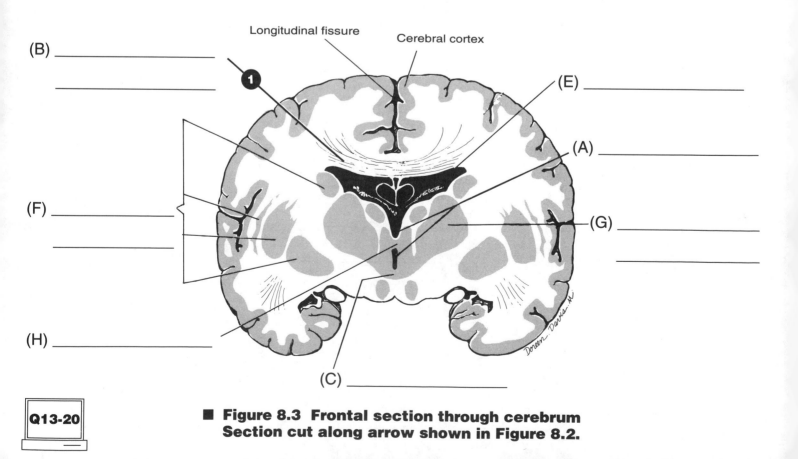

Longitudinal fissure   Cerebral cortex

(B) _____

_____

(E) _____

(A) _____

(F) _____

_____

(G) _____

_____

(H) _____

(C) _____

Q13-20

■ **Figure 8.3 Frontal section through cerebrum**
Section cut along arrow shown in Figure 8.2.

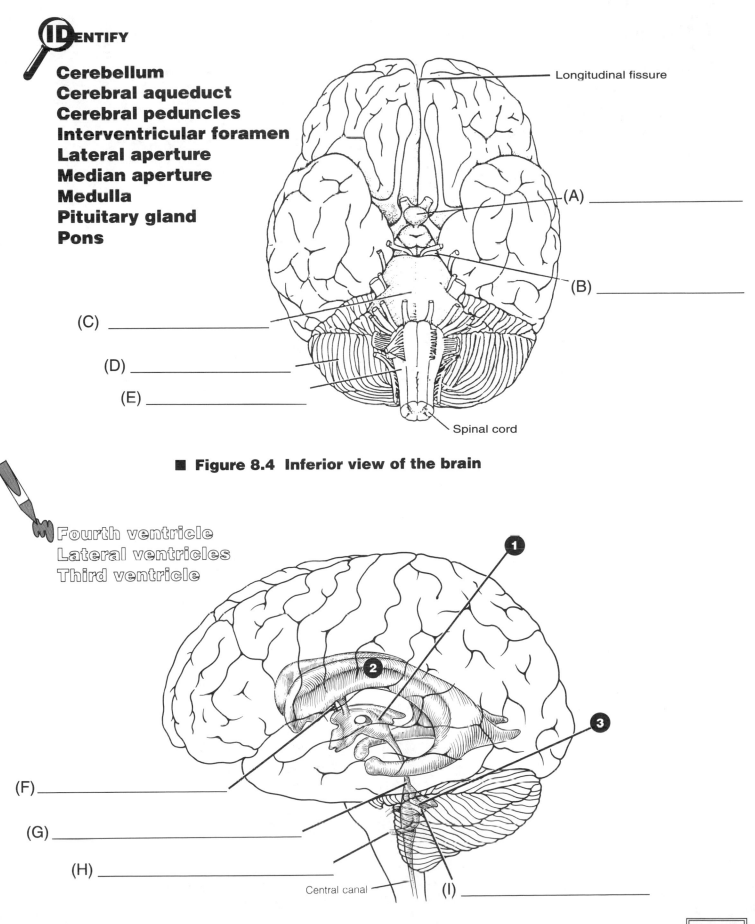

IDENTIFY

Cerebellum
Cerebral aqueduct
Cerebral peduncles
Interventricular foramen
Lateral aperture
Median aperture
Medulla
Pituitary gland
Pons

Longitudinal fissure

(A) _____

(B) _____

(C) _____

(D) _____

(E) _____

Spinal cord

■ **Figure 8.4 Inferior view of the brain**

Fourth ventricle
Lateral ventricles
Third ventricle

(F) _____

(G) _____

(H) _____

Central canal

(I) _____

Q21-28

■ **Figure 8.5 Internal impression of the ventricles**

ⅠⅮ ENTIFY

Arachnoid villi
Falx cerebri
Falx cerebelli
Superior sagittal sinus
Tentorium

Arachnoid
Dura mater
Pia mater
Subarachnoid space

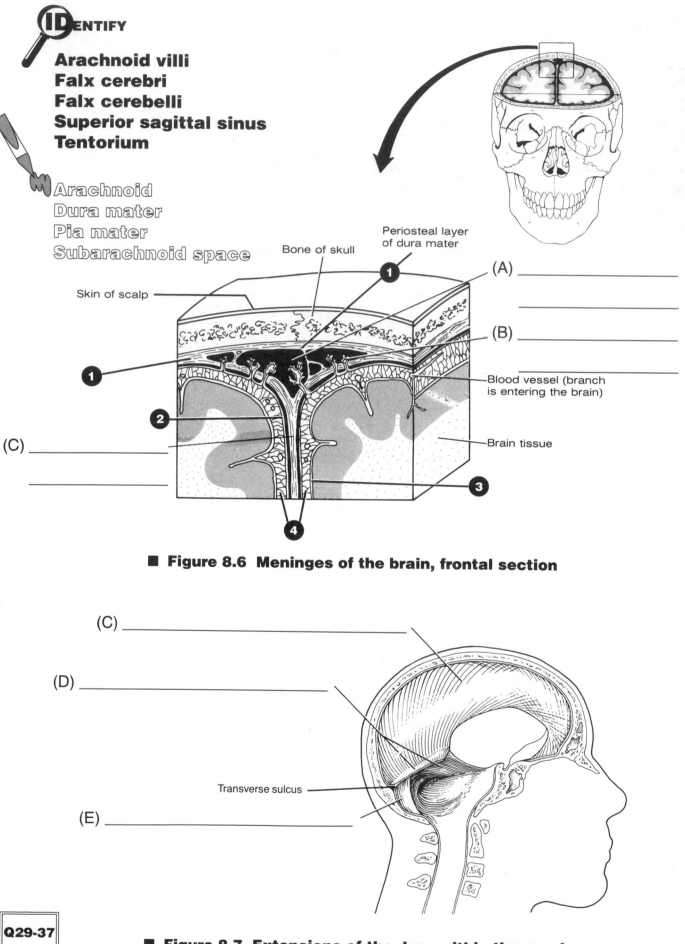

Bone of skull

Periosteal layer
of dura mater

Skin of scalp

(A) _____

_____

(B) _____

Blood vessel (branch
is entering the brain)

1

1

2

(C) _____

Brain tissue

3

_____

4

■ **Figure 8.6 Meninges of the brain, frontal section**

(C) _____

(D) _____

Transverse sulcus

(E) _____

■ **Figure 8.7 Extensions of the dura within the cranium**

IDENTIFY

Cauda equina
Conus medularis

Dura mater
Spinal cord

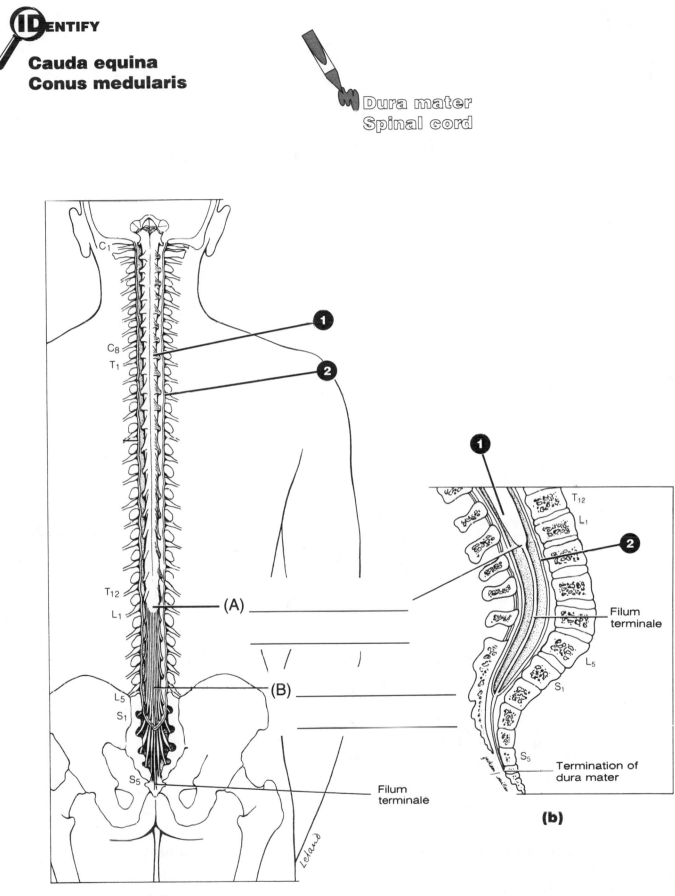

C₁

C₈
T₁

1

2

T₁₂
L₁
(A) _____

_____

(B) _____

L₅
S₁

1

T₁₂
L₁

2

Filum
terminale

L₅

S₁

S₅
Filum
terminale

S₅
Termination of
dura mater

(b)

(a)

■ **Figure 8.8 Spinal cord in (a) dorsal and (b) lateral views**

Q38-39

IDENTIFY

**Anterior median
sulcus
Central canal
Dorsal root
Dorsal root
ganglion
Posterior
median sulcus
Ventral root**

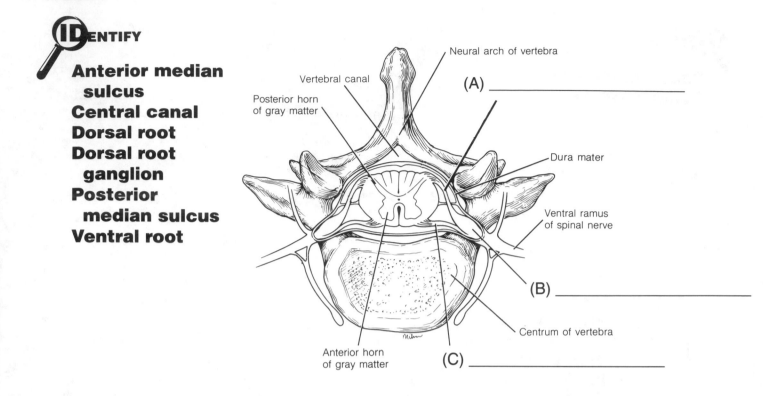

Neural arch of vertebra

Vertebral canal

Posterior horn
of gray matter

(A) _____

Dura mater

Ventral ramus
of spinal nerve

(B) _____

Centrum of vertebra

Anterior horn
of gray matter

(C) _____

■ **Figure 8.9 Spinal cord and vertebra, transverse view**

**Arachnoid
Dura mater
Pia mater**

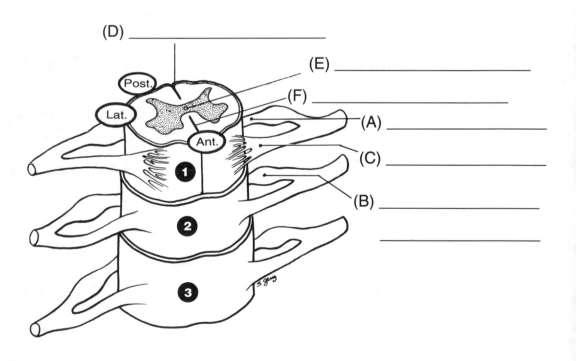

(D) _____

(E) _____

(F) _____

(A) _____

(C) _____

(B) _____

_____

Post.

Lat.

Ant.

■ **Figure 8.10 Structure of the spinal cord and meninges**

■ Table 8.2 Summary of Spinal White Matter Tracts

Fasciculus	Position P=posterior L=lateral A=anterior	Point of crossover E=entry M=medulla or above N=not crossed	Description S=sensory M=motor	Function
Fasciculus gracilis	(a)	(b)	S	Proprioception and fine touch of lower limbs and trunk
Fasciculus cuneatus	(c)	(d)	S	Proprioception and fine touch of upper body
Lateral spinothalamic	(e)	(f)	S	Pain and temperature sense
Ventral spinothalamic	(g)	(h)	S	Crude touch and pressure sense
Dorsal spinocerebellar	(i)	(j)	S	Unconscious proprioception
Ventral spinocerebellar	(k)	(l)	S	Unconscious proprioception
Pyramidal				
Lateral corticospinal	(m)	(n)	M	Motor impulse transmission from cerebrum to skeletal muscles on the opposite side of the body
Ventral corticospinal	(o)	(p)	M	Voluntary motor activity on the same side of the body
Extrapyramidal				
Rubrospinal	(q)	(r)	M	Motor impulse transmission for muscle tone and posture on opposite side of the body
Vestibulospinal	(s)	(t)	M	Head movements to maintain balance
Tectospinal	(u)	(v)	M	Head movements to visual/auditory stimulis
Olivospinal	(w)	(x)	M	Maintenance of muscle tone and posture

Q44-45

MODULE 9

The Peripheral Nervous System

IDENTIFY

Abducens VI
Facial VII
Glossopharyngeal IX
Hypoglossal XII

Oculomotor III
Olfactory I
Optic II
(Spino)Accessory XI

Trigeminal V
Trochlear IV
Vagus X
Vestibulocochlear VIII

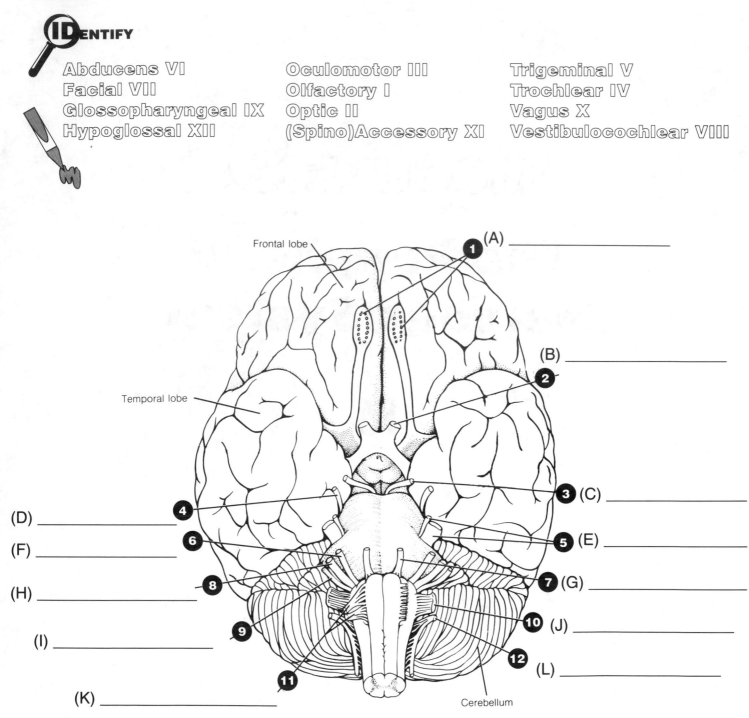

Frontal lobe

(A) _____

(B) _____

Temporal lobe

(C) _____

(D) _____

(E) _____

(F) _____

(G) _____

(H) _____

(J) _____

(I) _____

(L) _____

(K) _____

Cerebellum

■ **Figure 9.1 Basal view of the brain showing cranial nerve origins**

Q1-15

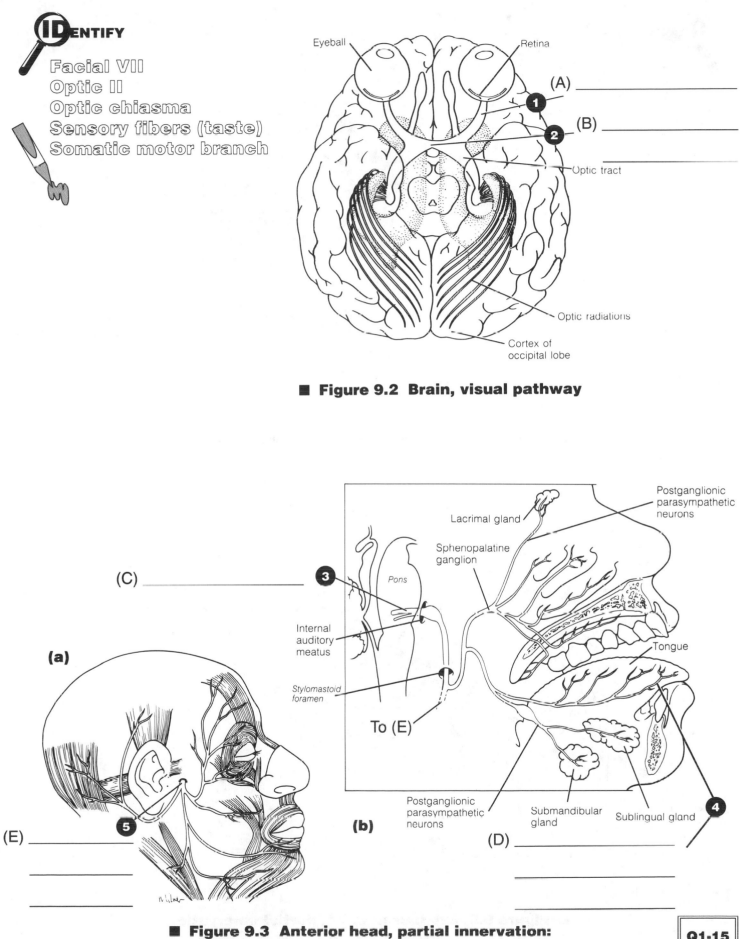

IDENTIFY

Facial VII
Optic II
Optic chiasma
Sensory fibers (taste)
Somatic motor branch

Eyeball
Retina
(A)
1
2
(B)
Optic tract
Optic radiations
Cortex of occipital lobe

■ **Figure 9.2 Brain, visual pathway**

(C)
3
Pons
Lacrimal gland
Postganglionic parasympathetic neurons
Sphenopalatine ganglion
Internal auditory meatus
Stylomastoid foramen
To (E)
Tongue
(a)
(E)
5
(b)
Postganglionic parasympathetic neurons
Submandibular gland
Sublingual gland
4
(D)

■ **Figure 9.3 Anterior head, partial innervation: (a) superficial and (b) deep views**

Q1-15

IDENTIFY

Abducens VI
Hypoglossal XII
Hypoglossal canal
Lateral rectus muscle

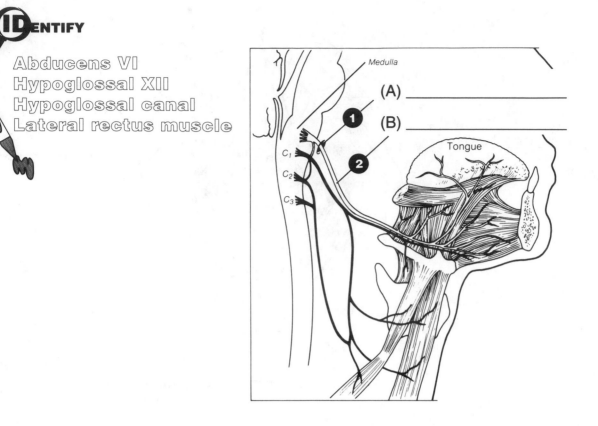

Medulla

(A) _____

(B) _____

Tongue

C_1
C_2
C_3

1
2

■ **Figure 9.4 Tongue musculature innervation**

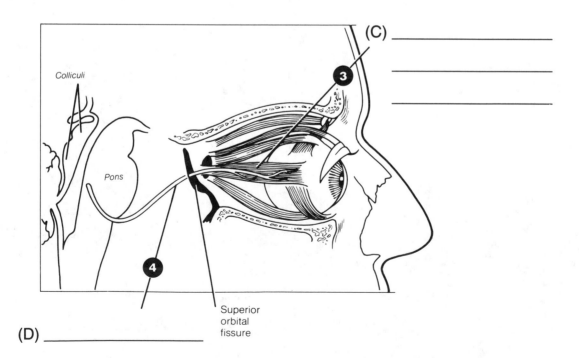

Colliculi

Pons

(C) _____

3

4

Superior
orbital
fissure

(D) _____

■ **Figure 9.5 Eye musculature, partial innervation**

Q1-15

IDENTIFY

Cranial root
Olfactory I
Olfactory bulb
Olfactory tract
Spinal root
(Spino)accessory XI

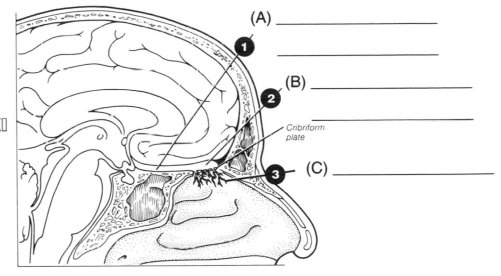

(A) _____

_____

(B) _____

_____

Cribriform plate

(C) _____

■ **Figure 9.6 Nasal mucosa, partial innervation**

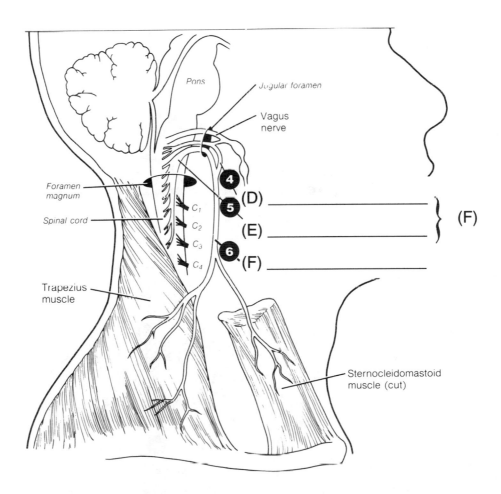

Pons

Jugular foramen

Vagus nerve

Foramen magnum

Spinal cord

C_1

C_2

C_3

C_4

Trapezius muscle

(D) _____

(E) _____ } (F)

(F) _____

Sternocleidomastoid muscle (cut)

■ **Figure 9.7 Neck musculature, partial innervation**

Q1-15

IDENTIFY

Cochlear branch
Vagus X
Vestibular branch
Vestibulocochlear VIII

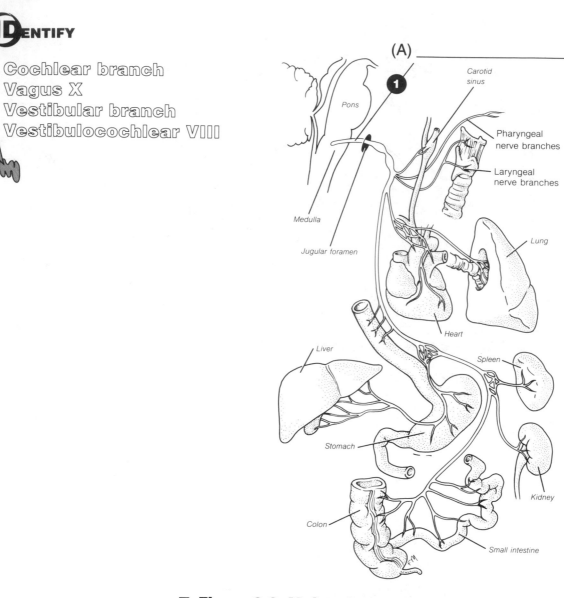

(A) _____

(1)

Pons

Carotid sinus

Pharyngeal nerve branches

Laryngeal nerve branches

Medulla

Lung

Jugular foramen

Heart

Liver

Spleen

Stomach

Kidney

Colon

Small intestine

■ **Figure 9.8 Major viscera, partial innervation**

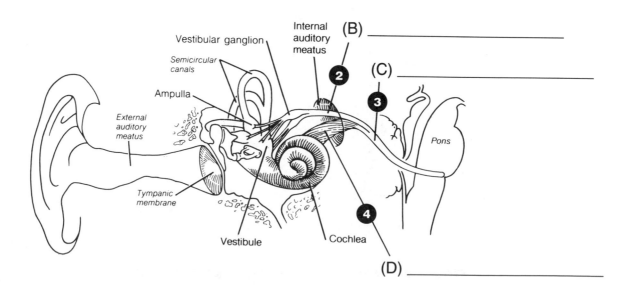

Vestibular ganglion

Semicircular canals

Internal auditory meatus

(B) _____

Ampulla

(2)

(C) _____

(3)

External auditory meatus

Pons

Tympanic membrane

(4)

Vestibule

Cochlea

(D) _____

Q1-15

■ **Figure 9.9 Inner ear innervation**

IDENTIFY

Extrinsic eye muscles
Glossopharyngeal IX
Oculomotor III
Parotid gland
Sensory fibers (taste)

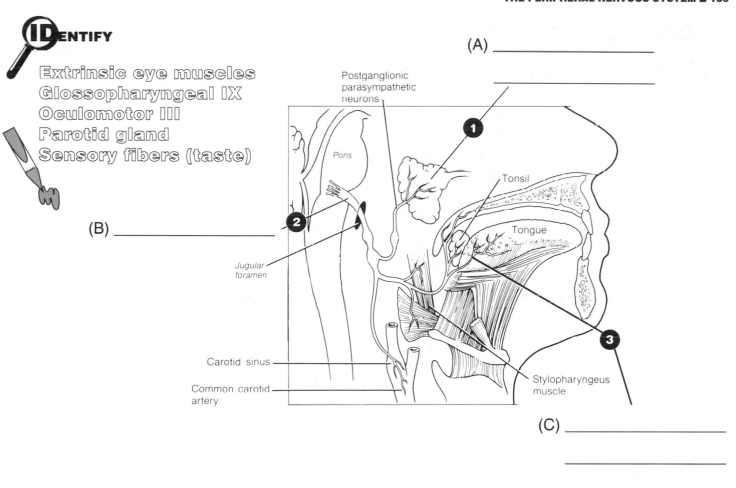

(A) _____

_____

Postganglionic
parasympathetic
neurons

Pons

1

Tonsil

Tongue

(B) _____

2

Jugular
foramen

3

Carotid sinus

Common carotid
artery

Stylopharyngeus
muscle

(C) _____

_____

_____

■ **Figure 9.10 Tongue and pharynx, partial innervation**

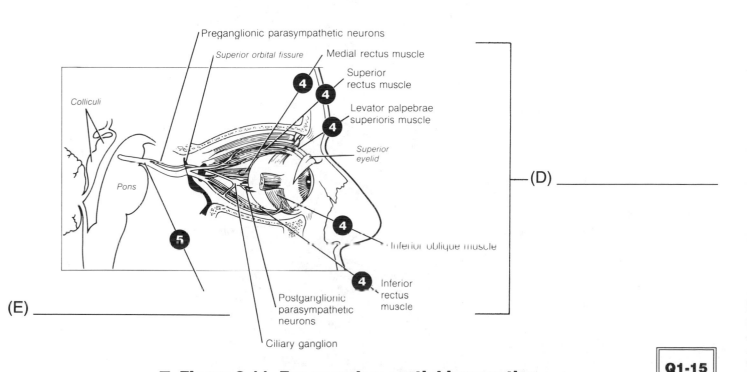

Preganglionic parasympathetic neurons

Superior orbital fissure   Medial rectus muscle

Superior
rectus muscle

Colliculi

4   4

Levator palpebrae
superioris muscle

4

Superior
eyelid

Pons

(D) _____

5

4

Inferior oblique muscle

4

Inferior
rectus
muscle

(E) _____

Postganglionic
parasympathetic
neurons

Ciliary ganglion

■ **Figure 9.11 Eye muscles, partial innervation**

Q1-15

IDENTIFY

Mandibular branch
Maxillary branch
Muscles of mastication
Ophthalmic branch
Superior oblique muscle
Trigeminal V
Trochlea
Trochlear IV

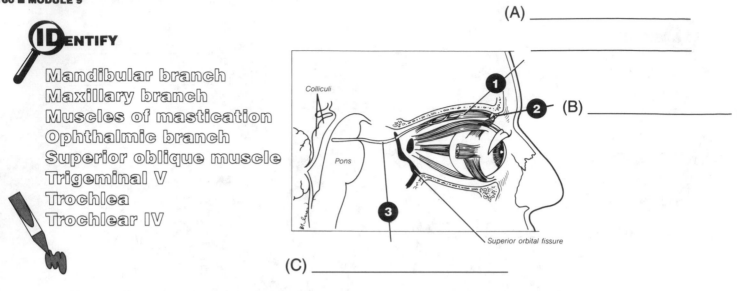

(A) _____

_____

(B) _____

(C) _____

■ **Figure 9.12 Extrinsic eye muscles, partial innervation**

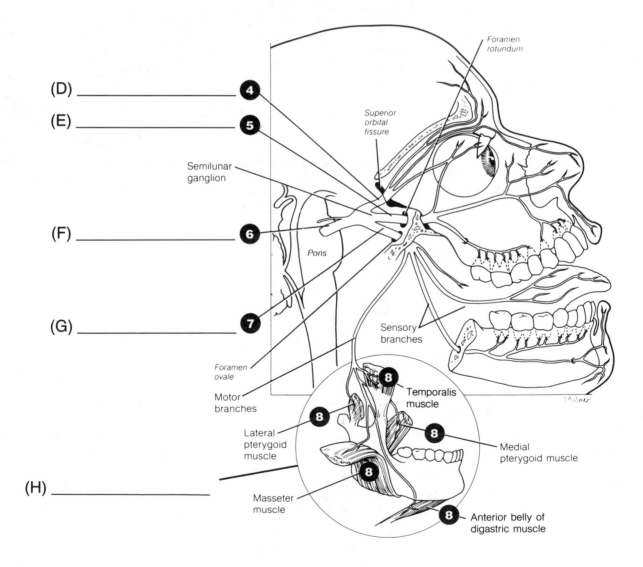

(D) _____

(E) _____

(F) _____

(G) _____

(H) _____

Q1-15

■ **Figure 9.13 Anterior head, partial innervation**

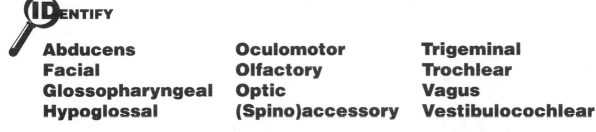

IDENTIFY

Abducens	Oculomotor	Trigeminal
Facial	Olfactory	Trochlear
Glossopharyngeal	Optic	Vagus
Hypoglossal	(Spino)accessory	Vestibulocochlear

■ **Table 9.1 Cranial Nerve Names and General Functions
(Diagrams of the cranial nerves are randomly arranged
in Figures 9.2–9.13)**

Cranial Nerve	Name	Sensory, Motor,* or Mixed	Regions or Organs Served	Figure Number
I	(A)	Sensory	Olfactory mucosa of nasal cavity; sense of smell	(B)
II	(C)	Sensory	Retina of eye; vision	(D)
III	(E)	Motor*	Eyelids; 4 of 6 extrinsic eye muscles; iris and lens	(F)
IV	(G)	Motor*	Superior oblique muscle of the eye	(H)
V	(I)	Mixed	Sensory innervation of face Muscles of mastication	(J)
VI	(K)	Motor*	Lateral rectus muscle of eye	(L)
VII	(M)	Mixed	Taste, anterior two-thirds of tongue; facial expression; saliva and tear glands	(N)
VIII	(O)	Sensory	Vestibular division: equilibrium Cochlear division: hearing	(P)
IX	(Q)	Mixed	Pharyngeal muscles Taste, posterior one-third of tongue	(R)
X	(S)	Mixed	Abdominal and thoracic viscera	(T)
XI	(U)	Motor*	Trapezius and sternocleidomastoid muscles	(V)
XII	(W)	Motor*	Intrinsic and extrinsic tongue muscles	(X)

*Cranial nerves described as "motor" contain proprioceptors which convey a sense of muscle position. Such nerves are technically "mixed" and, perhaps, should be thought of as "primarily motor."

Q1-15

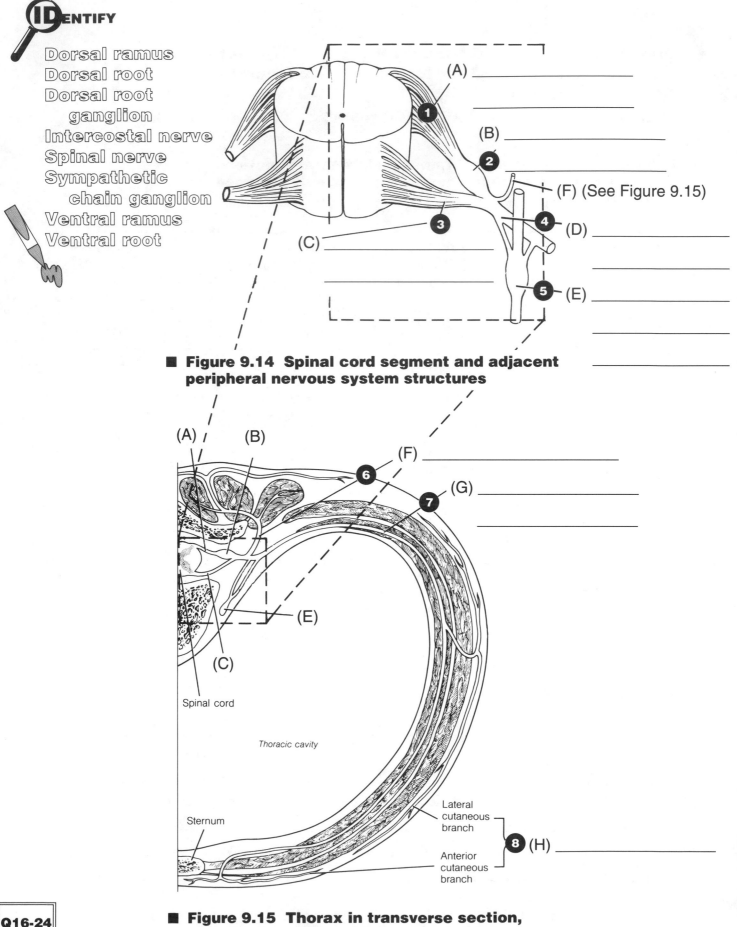

IDENTIFY

Dorsal ramus
Dorsal root
Dorsal root
  ganglion
Intercostal nerve
Spinal nerve
Sympathetic
  chain ganglion
Ventral ramus
Ventral root

(A) _____

_____

(B) _____

_____

(F) (See Figure 9.15)

(C) _____

(D) _____

_____

(E) _____

_____

_____

■ **Figure 9.14 Spinal cord segment and adjacent peripheral nervous system structures**

(A)   (B)

(F) _____

(G) _____

_____

(E)

(C)

Spinal cord

Thoracic cavity

Sternum

Lateral cutaneous branch

Anterior cutaneous branch

(H) _____

Q16-24

■ **Figure 9.15 Thorax in transverse section, showing formation of intercostal nerves**

IDENTIFY

Autonomic
Motor
Parasympathetic
Sensory
Somatic
Sympathetic

■ **Table 9.2 Functional Components of Spinal Nerves**
(Refer to Figures 9.14 and 9.15)

Terms Descriptive of Neuron Function	Meaning/Function	Distribution/ Location in PNS
(A)	Efferent, e.g., "outgoing" from CNS to PNS	Fibers comprise ventral root of all spinal nerves; extends to all glands and muscles
(B)	Involuntary; controls glands, smooth muscle, and cardiac muscle	Fibers pass through most spinal nerves; cell bodies in peripheral ganglia
(C)	Involuntary; generally maintains body under quiet conditions; "quiescence"	Fibers in cranial nerves III, VII, IX, X; those in ventral roots of sacral nerves form pelvic splanchnic plexus
(D)	Involuntary; generally prepares body for emergency conditions; "arousal"	Fibers in most spinal nerves; paravertebral ganglia parallel to vertebral column contain nuclei of postganglionic neurons
(E)	Voluntary; innervates skeletal muscle only	Fibers in cranial nerves III, IV, V, VI, VII, IX, X, XI, XII and all spinal nerves
(F)	Afferent pathways to the CNS; provides conscious and subconscious sensation	Fibers in all cranial and spinal nerves; cell bodies in ganglia outside CNS, including dorsal root ganglia

Q16-21

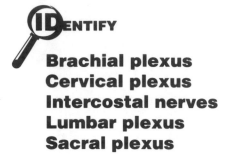

Brachial plexus
Cervical plexus
Intercostal nerves
Lumbar plexus
Sacral plexus

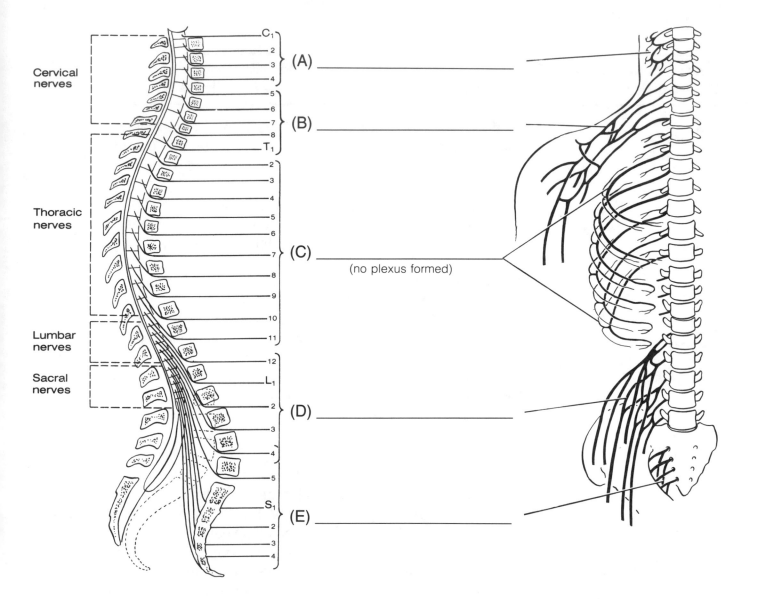

Cervical
nerves

Thoracic
nerves

Lumbar
nerves

Sacral
nerves

(A) _____

(B) _____

(C) _____
(no plexus formed)

(D) _____

(E) _____

■ **Figure 9.16 Plexus formation from the ventral
ramus of spinal nerves**

Q27

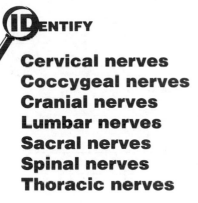

IDENTIFY

Cervical nerves
Coccygeal nerves
Cranial nerves
Lumbar nerves
Sacral nerves
Spinal nerves
Thoracic nerves

■ **Table 9.3 Naming of Peripheral Nervous System Structures**
(Refer to Figures 9.16 and 9.1)

Peripheral Nervous System Components	Names Applied	Naming Convention Observed
(A)	I-XII	By Roman numerals, from anterior to posterior; also, by area served
(B)	C_1–Cg_1	Somewhat variable, but with reference to vertebrae superior or inferior to the spinal nerve in question
(C)	C_1–C_8	Named for vertebrae above which they emerge, except spinal nerve C_8 emerges below vertebra C_7
(D)	T_1–T_{12}	Named for vertebrae below which they emerge
(E)	L_1–L_5	Named for vertebrae below which they emerge
(F)	S_1–S_5	Named for vertebrae below which they emerge
(G)	Cg_1	Emerges below the coccygeal vertebrae

Q1,
25-26

IDENTIFY

Axillary nerve
Median nerve
Musculocutaneous
  nerve
Phrenic nerve
Radial nerve
Ulnar nerve

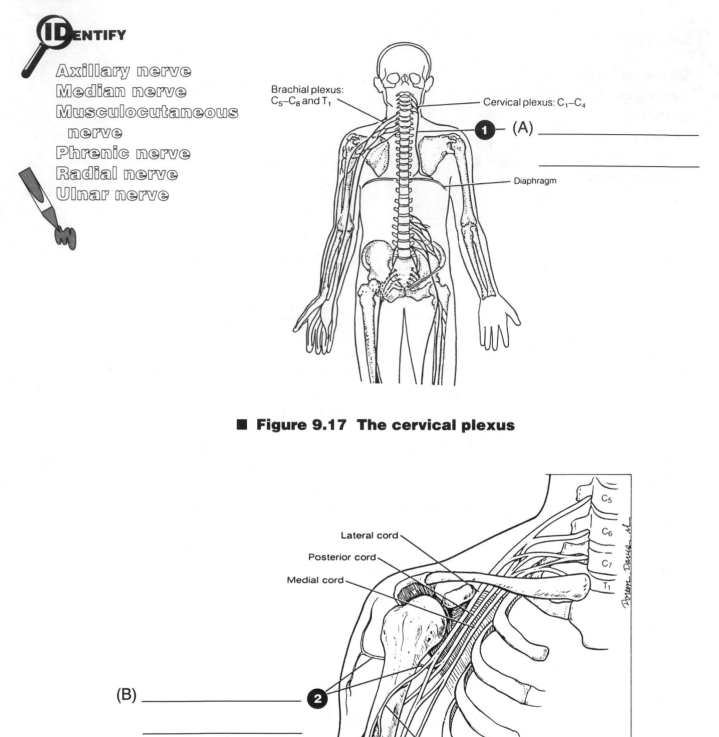

Brachial plexus:
C_5–C_8 and T_1

Cervical plexus: C_1–C_4

1 (A) _____

_____

Diaphragm

■ **Figure 9.17 The cervical plexus**

Lateral cord

Posterior cord

Medial cord

C_5

C_6

C_7

T_1

(B) _____

_____

(C) _____

2

3

4 (D) _____

_____

5 (E) _____

_____

6

(F) _____

_____

■ **Figure 9.18 The brachial plexus**

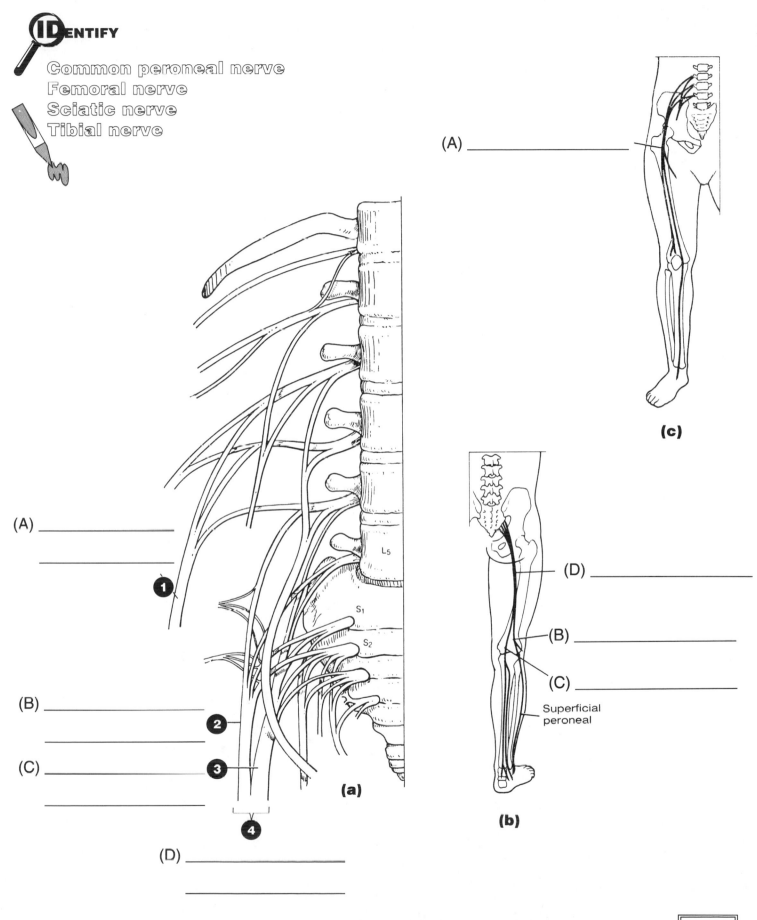

IDENTIFY

Common peroneal nerve
Femoral nerve
Sciatic nerve
Tibial nerve

(A) _____

(A) _____

_____

1

(B) _____

_____

(C) _____

2

3

4

(D) _____

_____

L₅

S₁

S₂

(a)

(D) _____

(B) _____

(C) _____

Superficial peroneal

(b)

(c)

■ **Figure 9.19 The lumbosacral plexus: (a) lumbosacral, (b) sacral, and (c) lumbar plexuses**

Q33-35

MODULE 10

The Special Senses

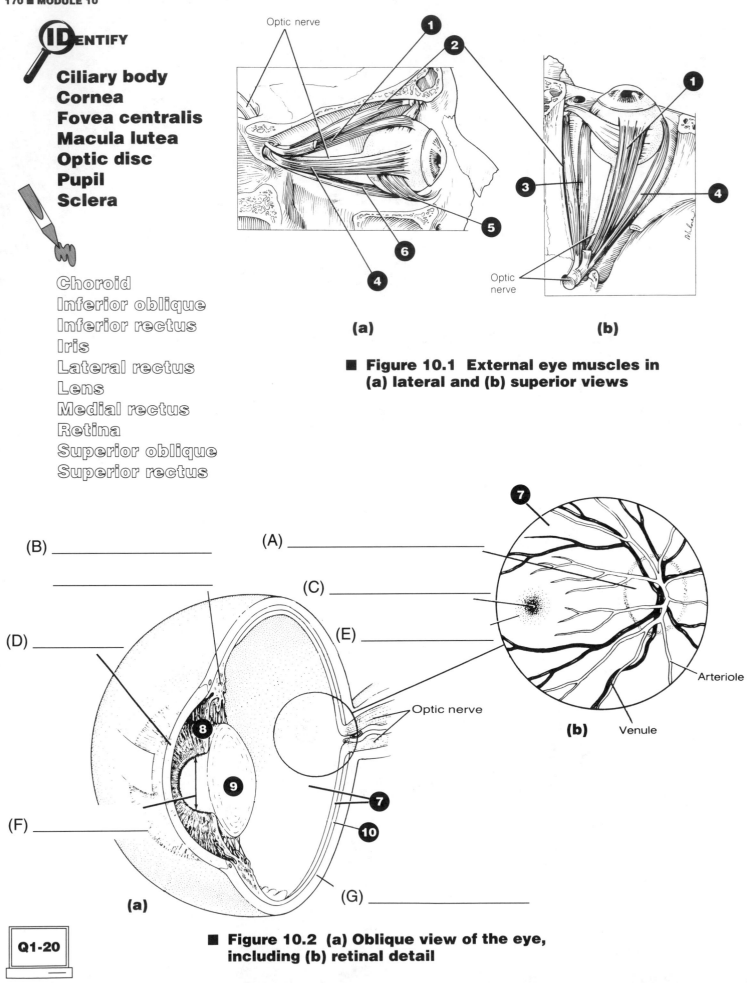

IDENTIFY

Ciliary body
Cornea
Fovea centralis
Macula lutea
Optic disc
Pupil
Sclera

Choroid
Inferior oblique
Inferior rectus
Iris
Lateral rectus
Lens
Medial rectus
Retina
Superior oblique
Superior rectus

Optic nerve

(a)   (b)

■ **Figure 10.1 External eye muscles in
(a) lateral and (b) superior views**

(B) _____

_____

(D) _____

(F) _____

(a)

(A) _____

(C) _____

(E) _____

Optic nerve

Arteriole

Venule

(b)

(G) _____

Q1-20

■ **Figure 10.2 (a) Oblique view of the eye,
including (b) retinal detail**

IDENTIFY

Anterior chamber
Canal of Schlemm
Posterior chamber
Sclera
Suspensory ligaments

Choroid
Ciliary muscles
Conjunctiva
Iris

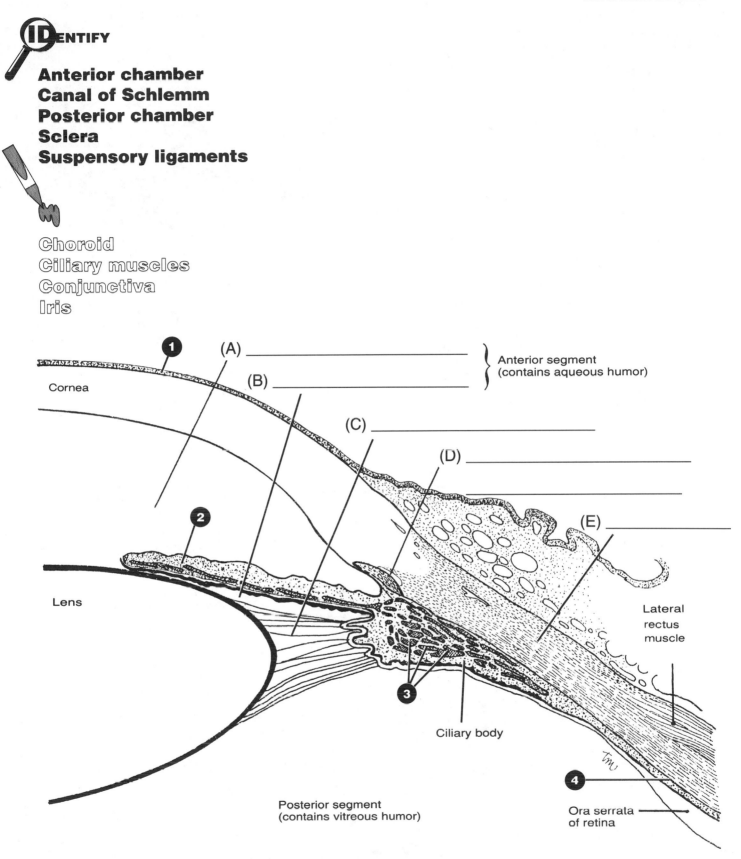

Cornea

Lens

(A) _____ } Anterior segment (contains aqueous humor)

(B) _____

(C) _____

(D) _____

(E) _____

Lateral rectus muscle

Ciliary body

Posterior segment (contains vitreous humor)

Ora serrata of retina

■ Figure 10.3 Ciliary body and adjacent structures

Q8-14
Q19-20

IDENTIFY

Ampulla
Auricle
External
 auditory
 meatus
Oval
 window
Round
 window
Saccule
Utricle

Auditory tube
Endolymph
Incus
Malleus
Perilymph
Stapes
Tympanic
 membrane

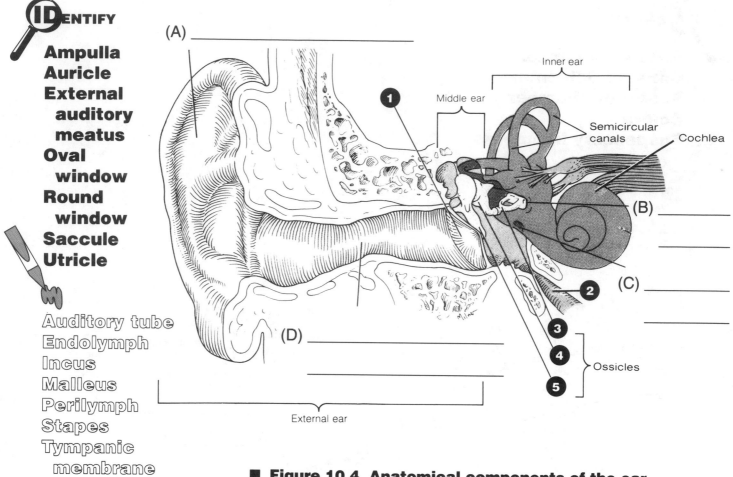

(A) _____

Inner ear

① Middle ear

Semicircular canals

Cochlea

(B) _____

(C) _____

_____

② ③ ④ ⑤ } Ossicles

(D) _____

External ear

■ **Figure 10.4 Anatomical components of the ear**

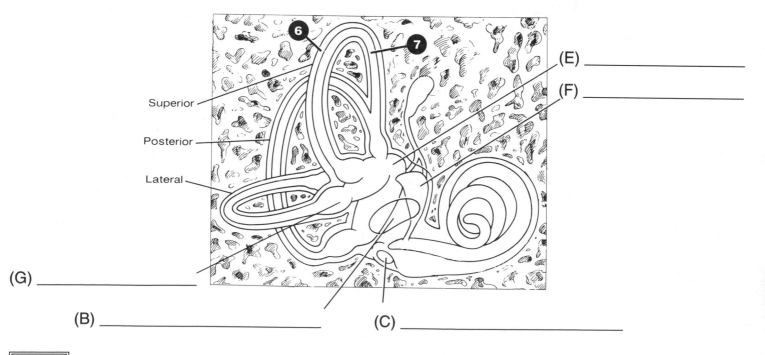

⑥ ⑦

Superior

Posterior

Lateral

(E) _____

(F) _____

(G) _____

(B) _____ (C) _____

■ **Figure 10.5 Membranous labyrinth suspended in the osseous labyrinth, greatly enlarged**

IDENTIFY

Basilar membrane
Cochlear nerve
Scala media (cochlear duct)
Scala tympani
Scala vestibuli
Tectorial membrane

Cochlear nerve
Endolymph
Perilymph

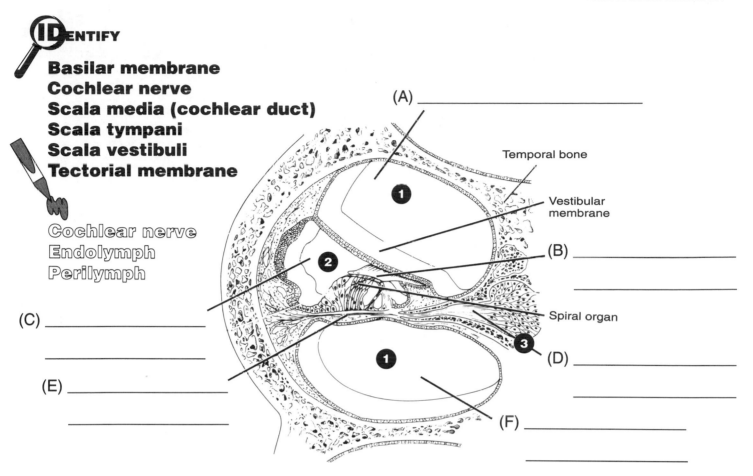

(A) _____

Temporal bone

Vestibular membrane

(B) _____

_____

Spiral organ

(C) _____

_____

(D) _____

(E) _____

_____

_____

(F) _____

_____

■ **Figure 10.6 Cross section through one turn of the cochlea**

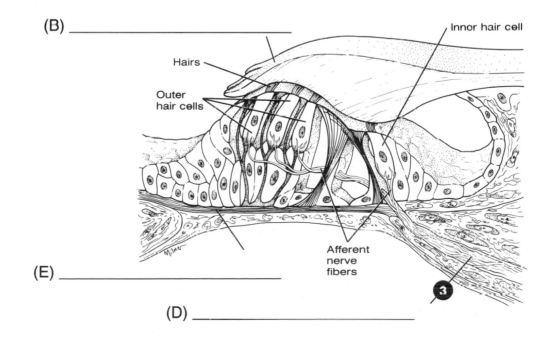

(B) _____

Hairs

Outer hair cells

Innor hair cell

Afferent nerve fibers

(E) _____

(D) _____

Q30-35

■ **Figure 10.7 Enlarged section through the spiral organ**

MODULE 11

The Endocrine System

IDENTIFY

Adenohypophysis (anterior lobe)
Adrenocorticotropin
Antidiuretic hormone
Follicle-stimulating hormone
Growth hormone
Luteinizing hormone
Neurohypophysis (posterior lobe)
Oxytocin
Prolactin
Thyrotropin

Adenohypophysis
(anterior lobe)
Hypothalamus
Neurohypophysis
(posterior lobe)

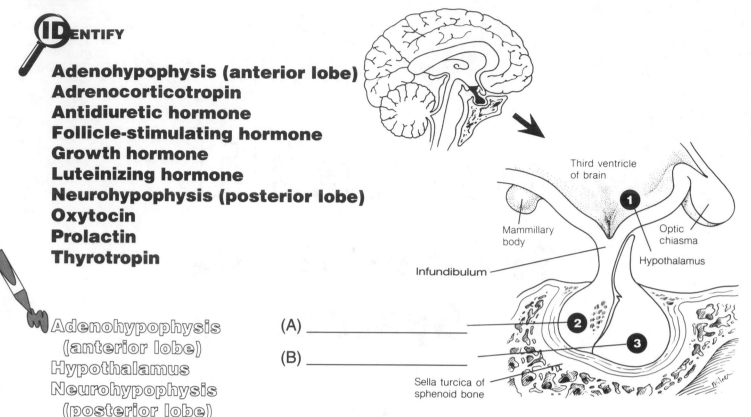

(A) _____

(B) _____

■ **Figure 11.1 The pituitary (hypophysis)**

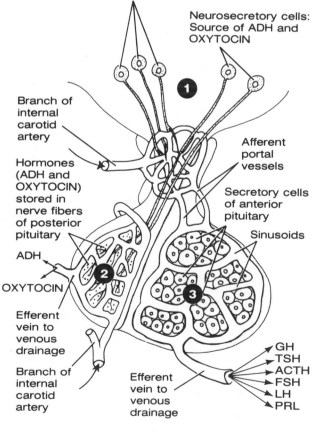

Hormones produced or released:

Anterior lobe

(C) (GH) _____

(D) (TSH) _____

(E) (ACTH) _____

(F) (FSH) _____

(G) (LH) _____

(H) (PRL) _____

Posterior lobe

(I) (ADH) _____

(J) _____

Q1-7

■ **Figure 11.2 Hypothalamus and pituitary in neural and vascular relationships**

IDENTIFY

Pineal gland
Melatonin
(Thyro)calcitonin
Thyroid gland
Thyroxin

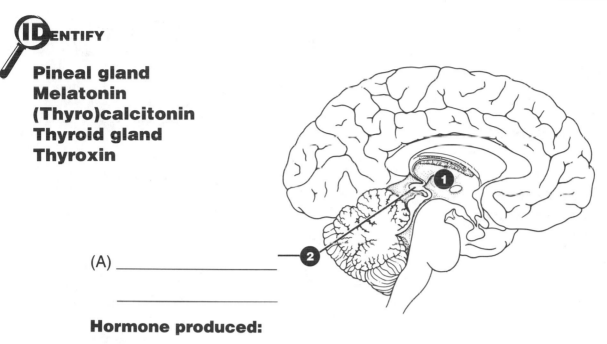

(A) _____

_____

Hormone produced:

(B) _____

■ **Figure 11.3 Midsagittal section through brain**

Larynx
Pineal gland
Third ventricle
Thyroid gland

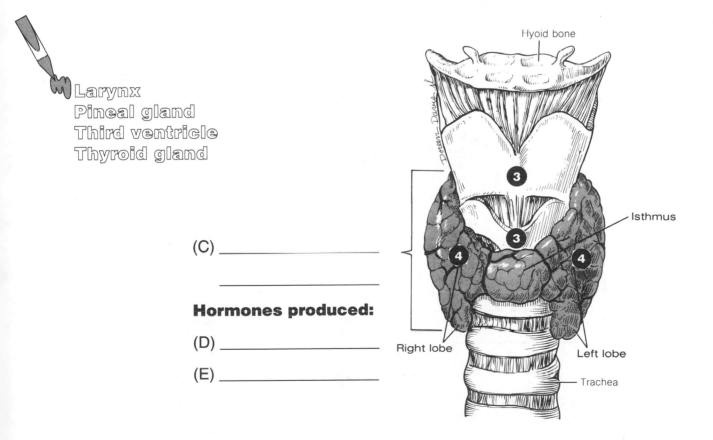

Hyoid bone

Isthmus

(C) _____

_____

Hormones produced:

(D) _____

(E) _____

Right lobe

Left lobe

Trachea

■ **Figure 11.4 Anterior view at larynx**

Q8-9

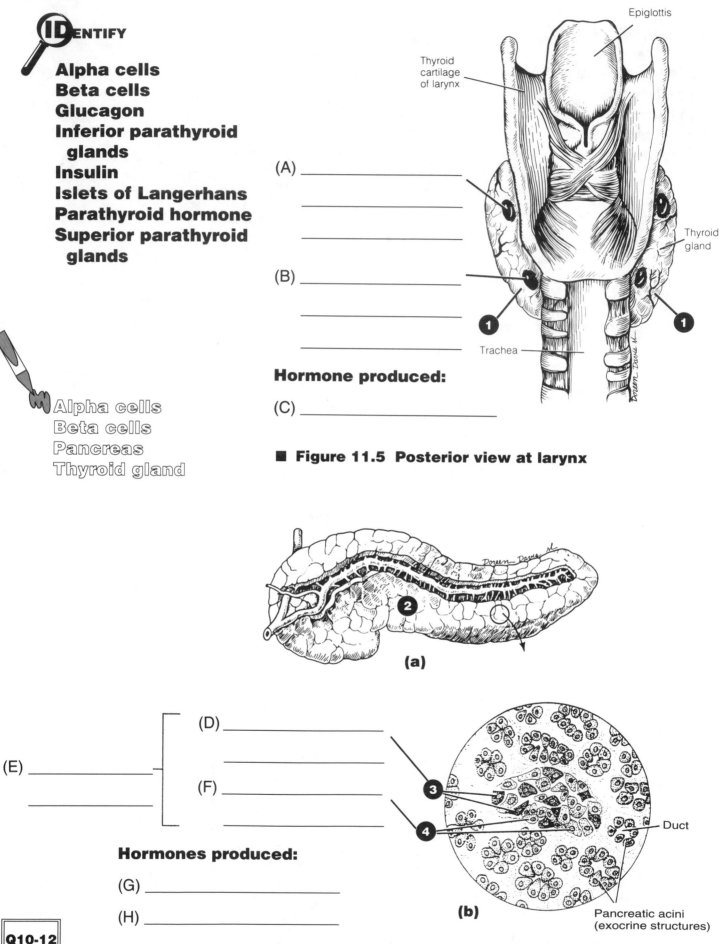

IDENTIFY

Alpha cells
Beta cells
Glucagon
Inferior parathyroid
glands
Insulin
Islets of Langerhans
Parathyroid hormone
Superior parathyroid
glands

Alpha cells
Beta cells
Pancreas
Thyroid gland

(A) _____

_____

_____

(B) _____

_____

_____

Hormone produced:

(C) _____

Epiglottis

Thyroid
cartilage
of larynx

Thyroid
gland

1   **1**

Trachea

■ **Figure 11.5 Posterior view at larynx**

2

(a)

(D) _____

(E) _____

_____

(F) _____

_____

Hormones produced:

(G) _____

(H) _____

3

4

Duct

Pancreatic acini
(exocrine structures)

(b)

Q10-12

■ **Figure 11.6 Pancreas in (a) gross and (b) microscopic views**

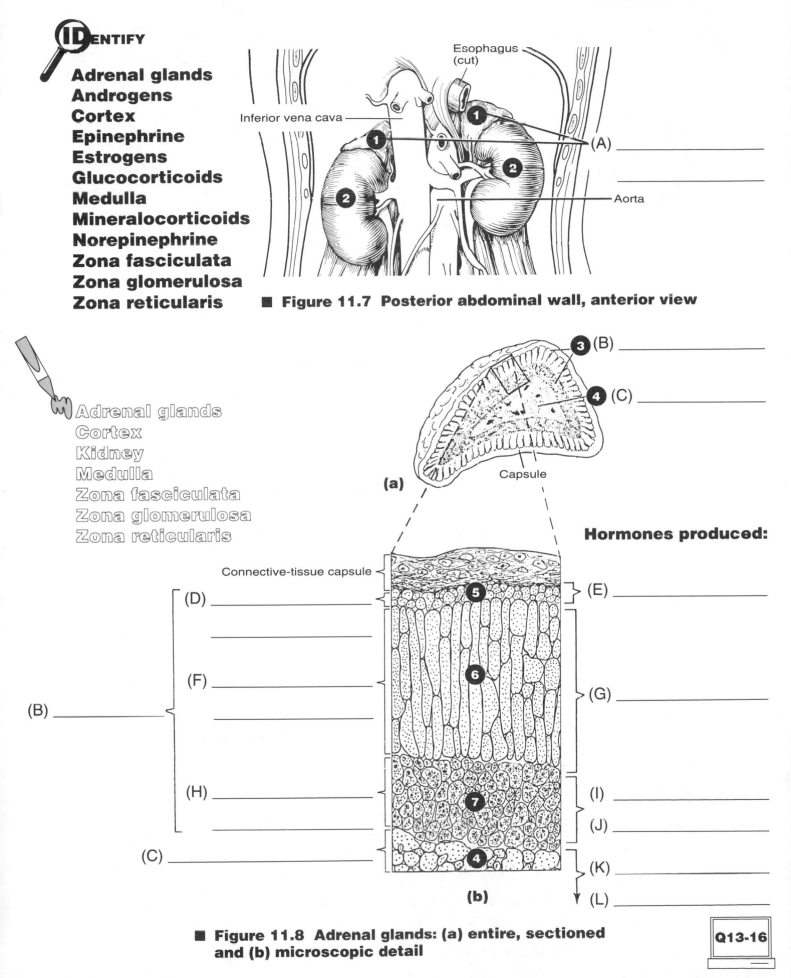

ⓘⒹENTIFY

Adrenal glands
Androgens
Cortex
Epinephrine
Estrogens
Glucocorticoids
Medulla
Mineralocorticoids
Norepinephrine
Zona fasciculata
Zona glomerulosa
Zona reticularis

Esophagus (cut)

Inferior vena cava

Aorta

■ Figure 11.7 Posterior abdominal wall, anterior view

Adrenal glands
Cortex
Kidney
Medulla
Zona fasciculata
Zona glomerulosa
Zona reticularis

(a)

Capsule

(B) _____

(C) _____

Hormones produced:

Connective-tissue capsule

(D) _____

(F) _____

(B) _____

(H) _____

(C) _____

(E) _____

(G) _____

(I) _____

(J) _____

(K) _____

(L) _____

(b)

■ Figure 11.8 Adrenal glands: (a) entire, sectioned and (b) microscopic detail

Q13-16

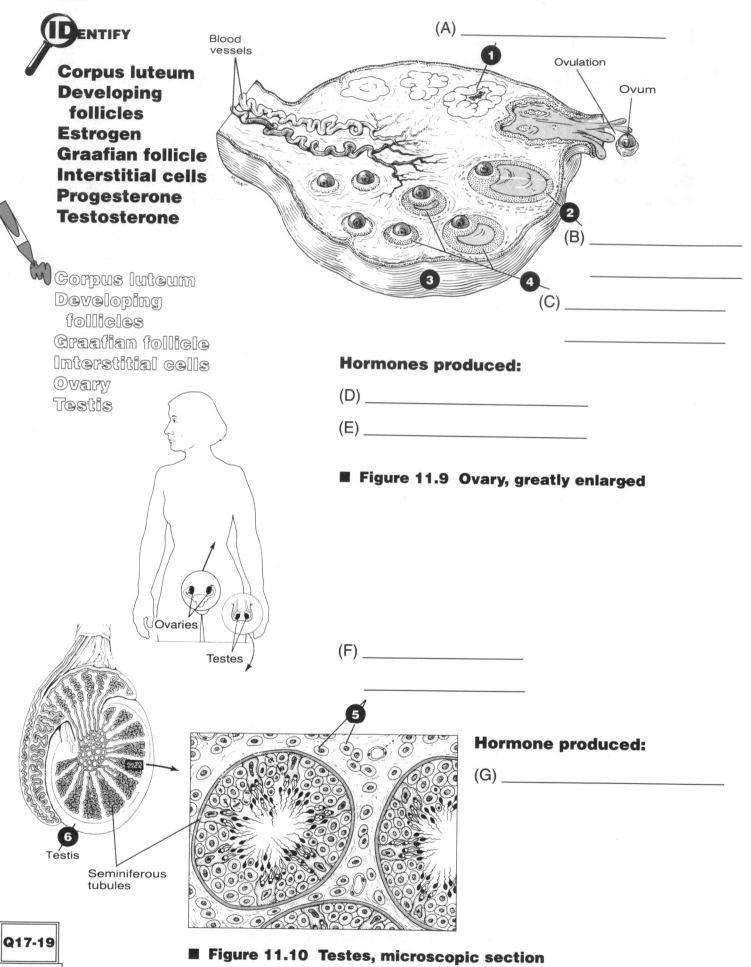

IDENTIFY

Corpus luteum
Developing
 follicles
Estrogen
Graafian follicle
Interstitial cells
Progesterone
Testosterone

Corpus luteum
Developing
 follicles
Graafian follicle
Interstitial cells
Ovary
Testis

Blood vessels

(A) _____

Ovulation

Ovum

(B) _____

_____

(C) _____

_____

Hormones produced:

(D) _____

(E) _____

■ **Figure 11.9 Ovary, greatly enlarged**

Ovaries

Testes

(F) _____

_____

Hormone produced:

(G) _____

Testis

Seminiferous tubules

Q17-19

■ **Figure 11.10 Testes, microscopic section**

IDENTIFY

Adrenal
Ovary
Pancreas
Parathyroid
Pineal
Pituitary
  (Hypophysis)
Testes
Thymus
Thyroid

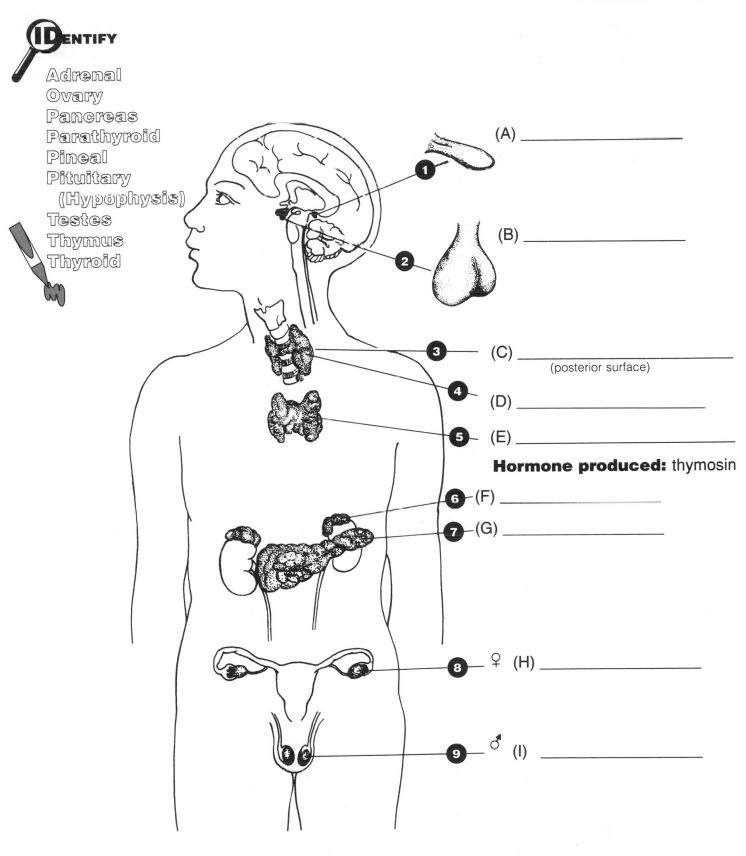

(A) _____

(B) _____

(C) _____
          (posterior surface)

(D) _____

(E) _____

Hormone produced: thymosin

(F) _____

(G) _____

♀ (H) _____

♂ (I) _____

Q20

■ **Figure 11.11 Relative locations of the endocrine glands**

MODULE 12

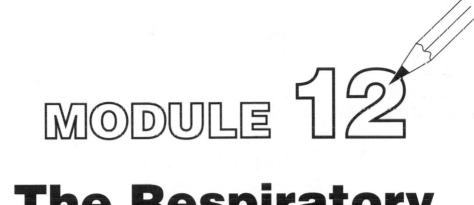

The Respiratory System

IDENTIFY

Auditory tube
External nares
Inferior meatus
Internal nares
Middle meatus
Superior meatus

(B) _____

_____

(C) _____

(E) _____

(F) _____

_____

Cricoid cartilage
Epiglottis
Hard palate
Inferior nasal concha
Lingual tonsil
Middle nasal concha
Palatine tonsil
Pharyngeal tonsil
Soft palate
Superior nasal concha
Thyroid cartilage

(A) _____

(D) _____

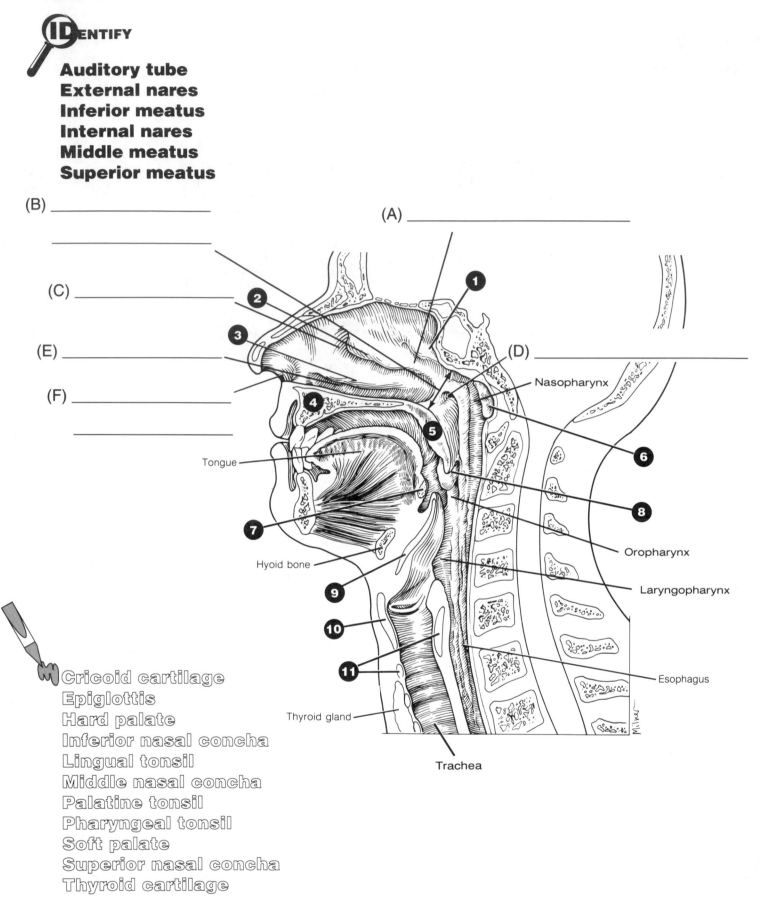

Nasopharynx

Tongue

Oropharynx

Hyoid bone

Laryngopharynx

Esophagus

Thyroid gland

Trachea

Q2-10

■ **Figure 12.1 Sagittal section through head and neck**

IDENTIFY

Glottis
True vocal cord
Ventricular fold
(Vestibular fold)

Arytenoid cartilage
Cricoid cartilage
Epiglottis
Thyroid cartilage
True vocal cord
Ventricular fold

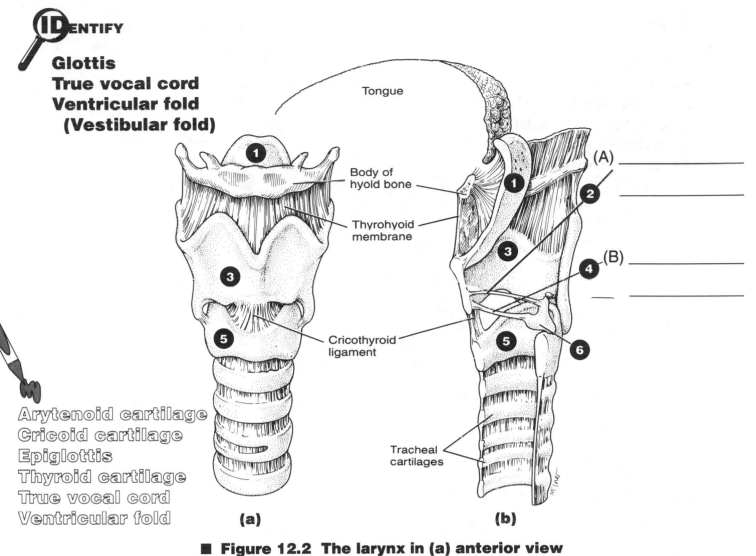

Tongue

Body of hyoid bone

Thyrohyoid membrane

Cricothyroid ligament

Tracheal cartilages

(A)

(B)

(a)   (b)

■ **Figure 12.2 The larynx in (a) anterior view**
and (b) sagittal section

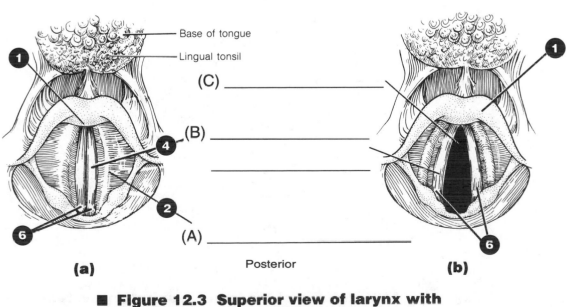

Anterior

Base of tongue

Lingual tonsil

(C) _____

(B) _____

_____

(A) _____

(a)   Posterior   (b)

■ **Figure 12.3 Superior view of larynx with**
(a) glottis closed and (b) glottis open

Q10-15

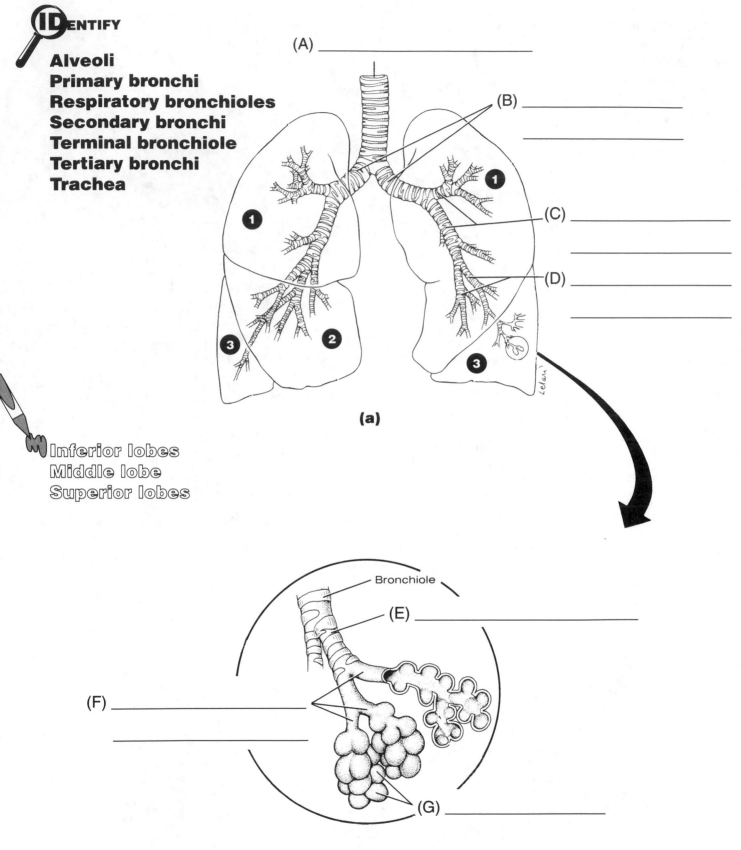

IDENTIFY

Alveoli
Primary bronchi
Respiratory bronchioles
Secondary bronchi
Terminal bronchiole
Tertiary bronchi
Trachea

(A) _____

(B) _____

_____

(C) _____

_____

(D) _____

_____

(a)

Inferior lobes
Middle lobe
Superior lobes

Bronchiole

(E) _____

(F) _____

_____

(G) _____

(b)

Q16-26

■ **Figure 12.4 Structures of the lower respiratory system:**
(a) lungs, trachea, and bronchi; (b) microscopic enlargement

MODULE 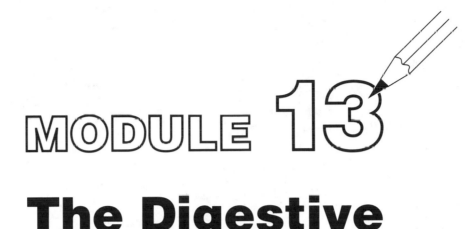 13

The Digestive System

IDENTIFY

Ascending colon   Esophagus   Pharynx
Cecum   Ileum   Rectum
Descending colon   Jejunum   Stomach
Duodenum   Oral cavity   Transverse colon

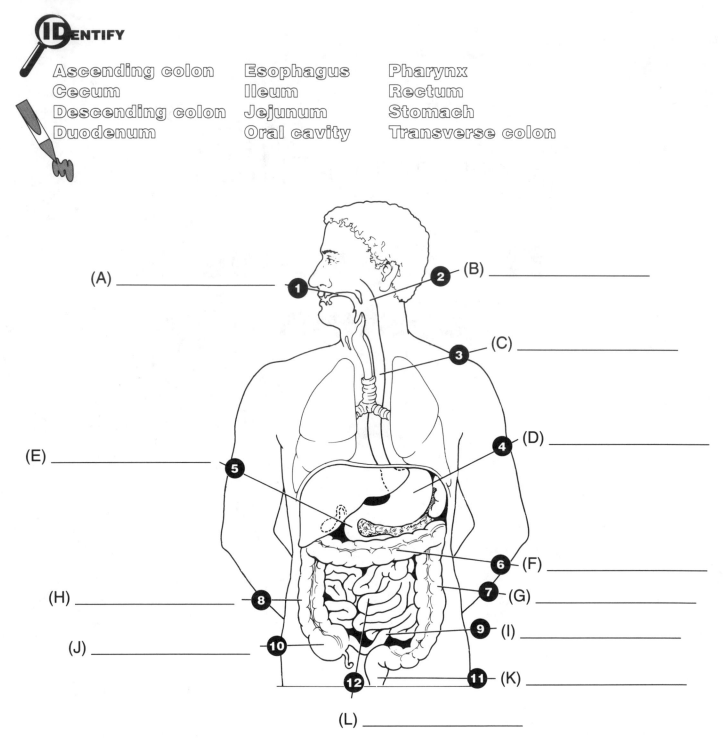

(A) _____

(B) _____

(C) _____

(D) _____

(E) _____

(F) _____

(G) _____

(H) _____

(I) _____

(J) _____

(K) _____

(L) _____

■ **Figure 13.1 Structures of the gastrointestinal tract**

Q1-24

IDENTIFY

Canines (cuspids)   Lingual frenulum   Oropharynx
Incisors   Molars (tricuspids) Premolars
Laryngopharynx   Oral cavity   Uvula

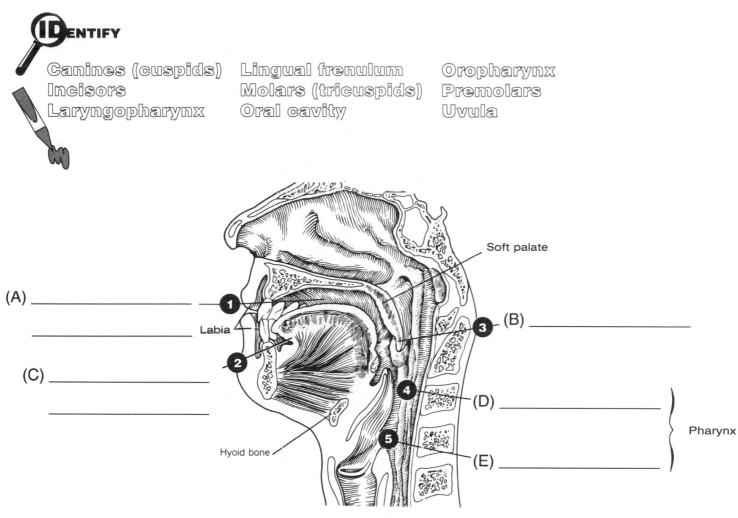

■ **Figure 13.2 Anterior-inferior head and neck, sagittal section**

(A) _____

_____   Labia

(C) _____

_____

Soft palate

(B) _____

(D) _____ } Pharynx

(E) _____

Hyoid bone

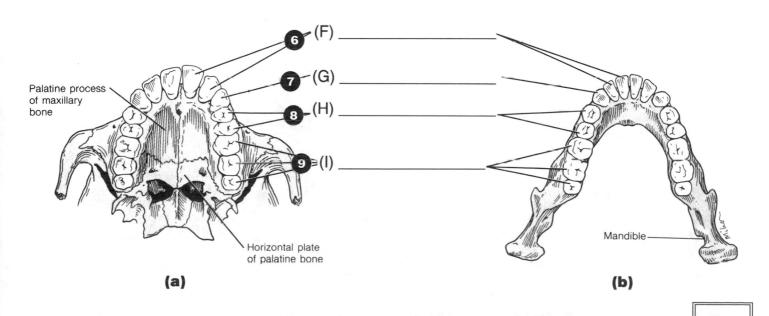

Palatine process
of maxillary
bone

(F) _____

(G) _____

(H) _____

(I) _____

Horizontal plate
of palatine bone

Mandible

(a)   (b)

■ **Figure 13.3 Permanent teeth of (a) upper jaw and (b) lower jaw**

Q1-4

IDENTIFY

Crown
Neck
Parotid gland
Root
Sublingual gland
Submandibular gland

(A) _____

(B) _____

(C) _____

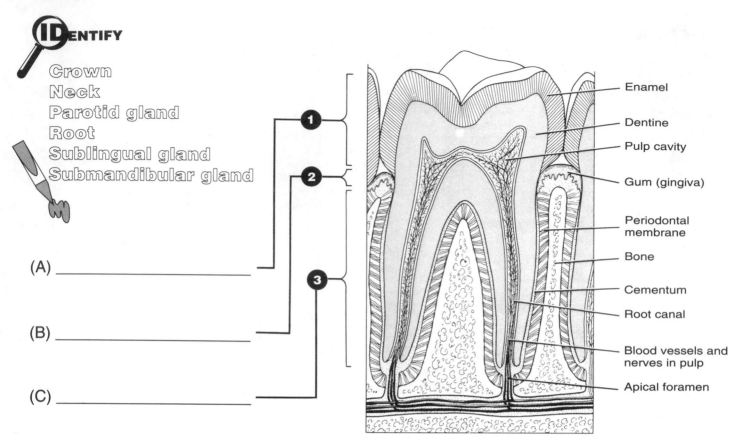

Enamel

Dentine

Pulp cavity

Gum (gingiva)

Periodontal membrane

Bone

Cementum

Root canal

Blood vessels and nerves in pulp

Apical foramen

■ Figure 13.4 Molar (tricuspid), vertical section

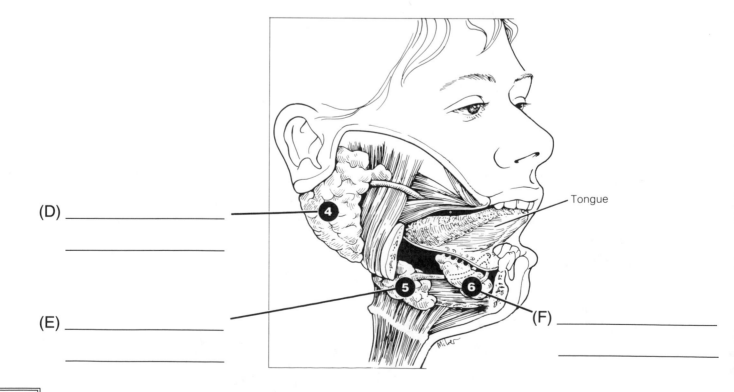

(D) _____

_____

(E) _____

Tongue

(F) _____

_____

Q5-7

■ Figure 13.5 The major salivary glands. The right side of the mandible has been removed.

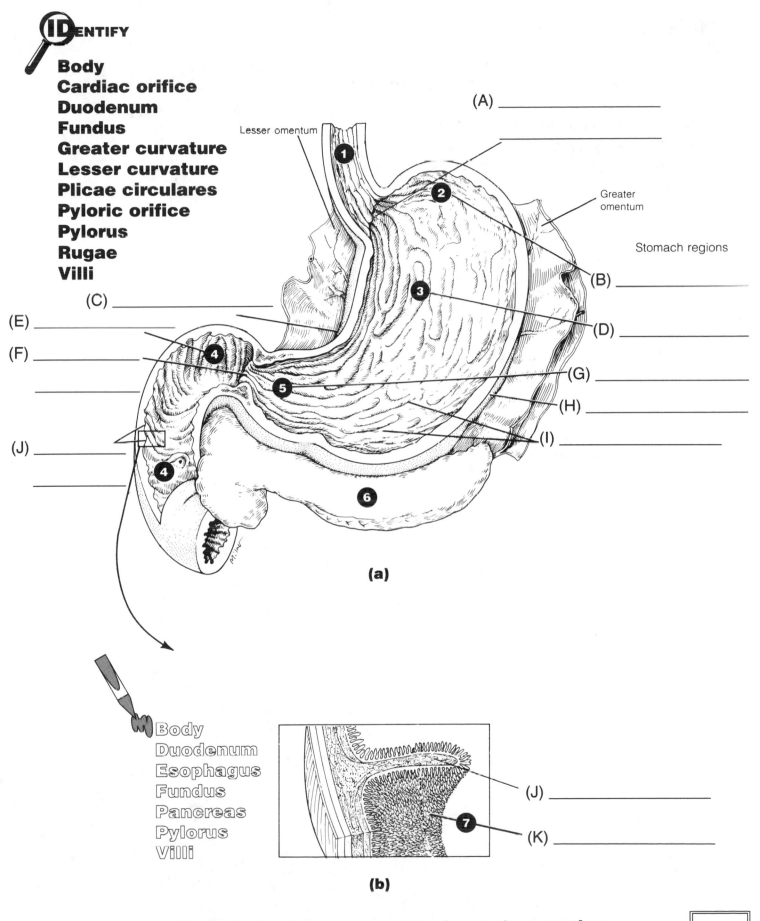

IDENTIFY

Body
Cardiac orifice
Duodenum
Fundus
Greater curvature
Lesser curvature
Plicae circulares
Pyloric orifice
Pylorus
Rugae
Villi

Lesser omentum

Greater omentum

Stomach regions

(A) _____

_____

(B) _____

(C) _____

(E) _____

(F) _____

_____

(D) _____

(G) _____

(H) _____

(I) _____

(J) _____

_____

(a)

Body
Duodenum
Esophagus
Fundus
Pancreas
Pylorus
Villi

(J) _____

(K) _____

(b)

■ **Figure 13.6 (a) Stomach and duodenum, long section;
(b) duodenum, detail**

ID ENTIFY

Common bile duct **Hepatic duct**
Coronary ligament **Hepatopancreatic**
Cystic duct **ampulla**
Duodenal papilla **Pancreas**
Falciform ligament **Pancreatic duct**
Gallbladder **Round ligament**

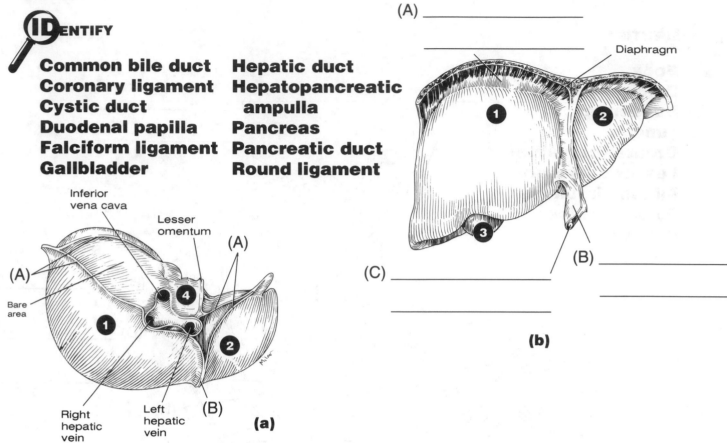

(A) _____

_____ Diaphragm

(B) _____

(C) _____

_____

(b)

Inferior
vena cava

Lesser
omentum

(A) _____

(A)

Bare
area

Right
hepatic
vein

Left
hepatic
vein

(B)

(a)

■ Figure 13.7 The liver in (a) superior and (b) anterior views

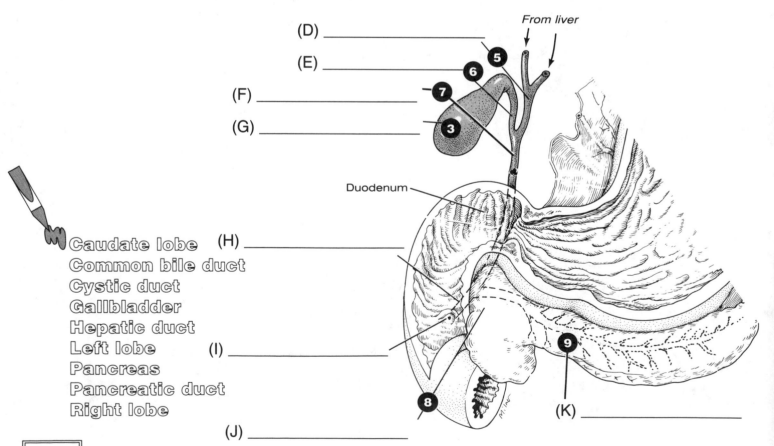

(D) _____

(E) _____

(F) _____

(G) _____

From liver

Duodenum

Caudate lobe (H) _____
Common bile duct
Cystic duct
Gallbladder
Hepatic duct
Left lobe (I) _____
Pancreas
Pancreatic duct
Right lobe

(J) _____

(K) _____

Q15-21

■ Figure 13.8 Exocrine ducts entering the duodenum

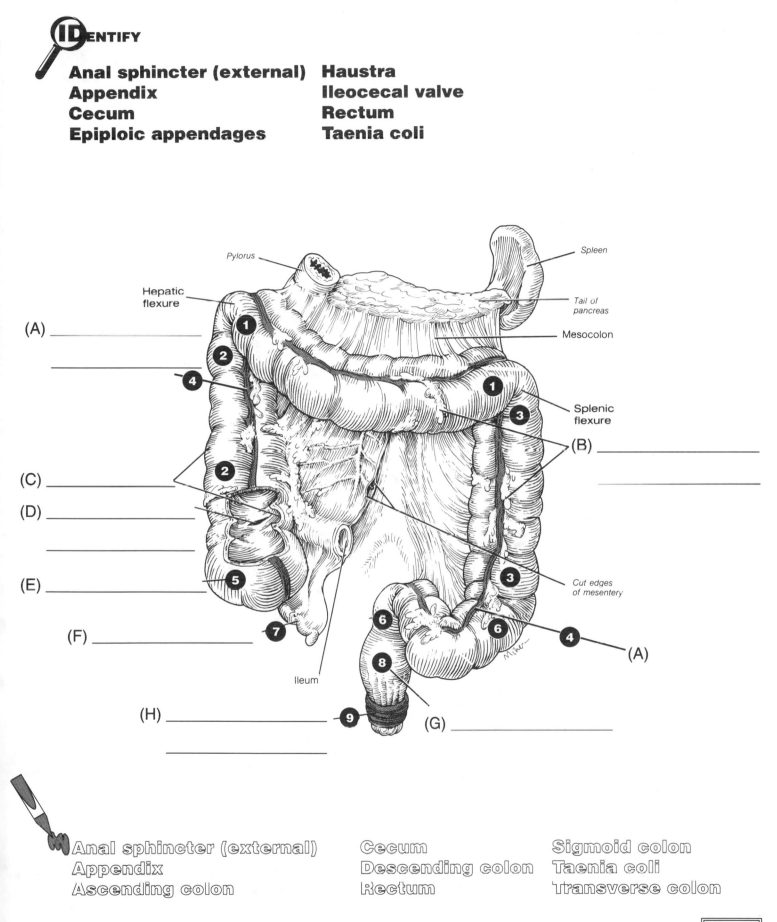

IDENTIFY

Anal sphincter (external) Haustra
Appendix Ileocecal valve
Cecum Rectum
Epiploic appendages Taenia coli

Pylorus

Spleen

Hepatic flexure

Tail of pancreas

Mesocolon

(A) _____

Splenic flexure

(B) _____

(C) _____

(D) _____

(E) _____

(F) _____

Cut edges of mesentery

Ileum

(H) _____

(G) _____

Anal sphincter (external) Cecum Sigmoid colon
Appendix Descending colon Taenia coli
Ascending colon Rectum Transverse colon

■ **Figure 13.9 The large intestine**

IDENTIFY

Coronary ligament   Mesocolon
Greater omentum   Parietal peritoneum
Intestinal mesenteries   Visceral peritoneum
Lesser omentum

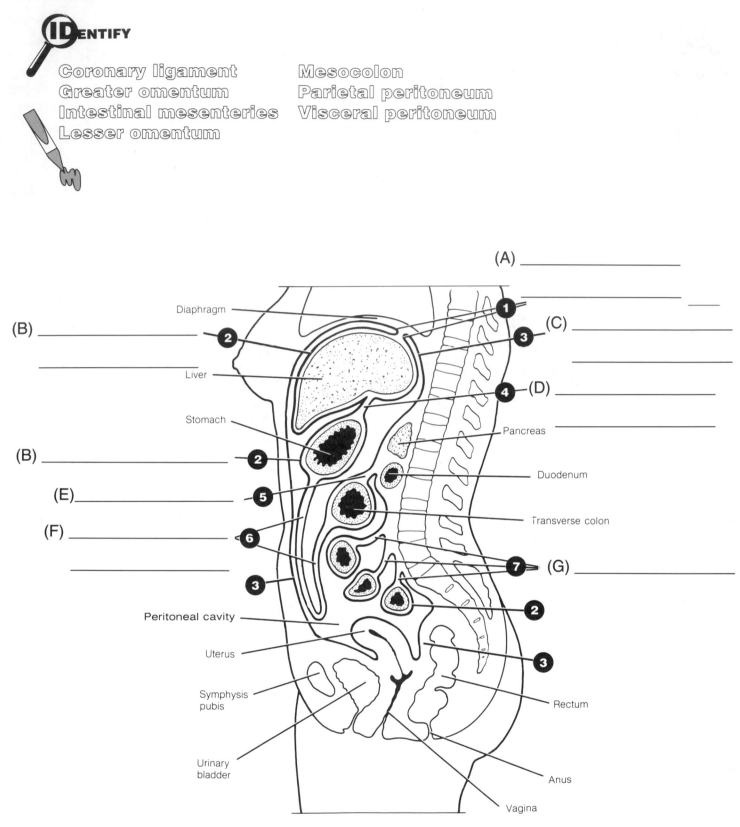

■ Figure 13.10 Peritoneal attachments, sagittal section

Q28-30

IDENTIFY

Falciform ligament
Greater omentum

(A) _____

_____

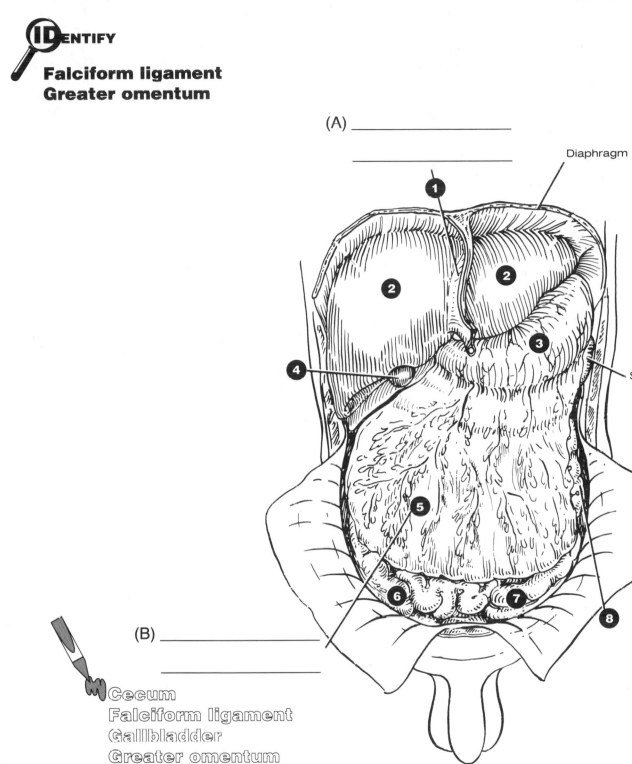

Diaphragm

Spleen

(B) _____

_____

Cecum
Falciform ligament
Gallbladder
Greater omentum
Large intestine
Liver
Small intestine
Stomach

■ **Figure 13.11 Abdominal viscera, superficial view**

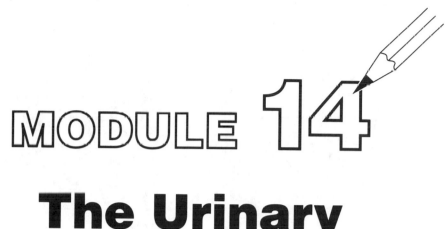

MODULE 14

The Urinary System

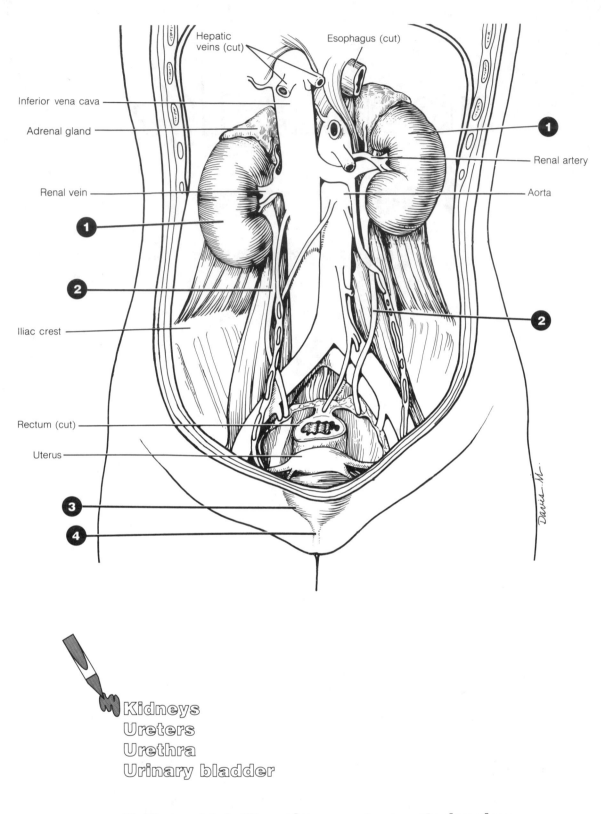

Hepatic veins (cut)

Esophagus (cut)

Inferior vena cava

Adrenal gland

Renal vein

1

1

Renal artery

Aorta

2

Iliac crest

2

Rectum (cut)

Uterus

3

4

Kidneys
Ureters
Urethra
Urinary bladder

■ Figure 14.1 The urinary system, anterior view

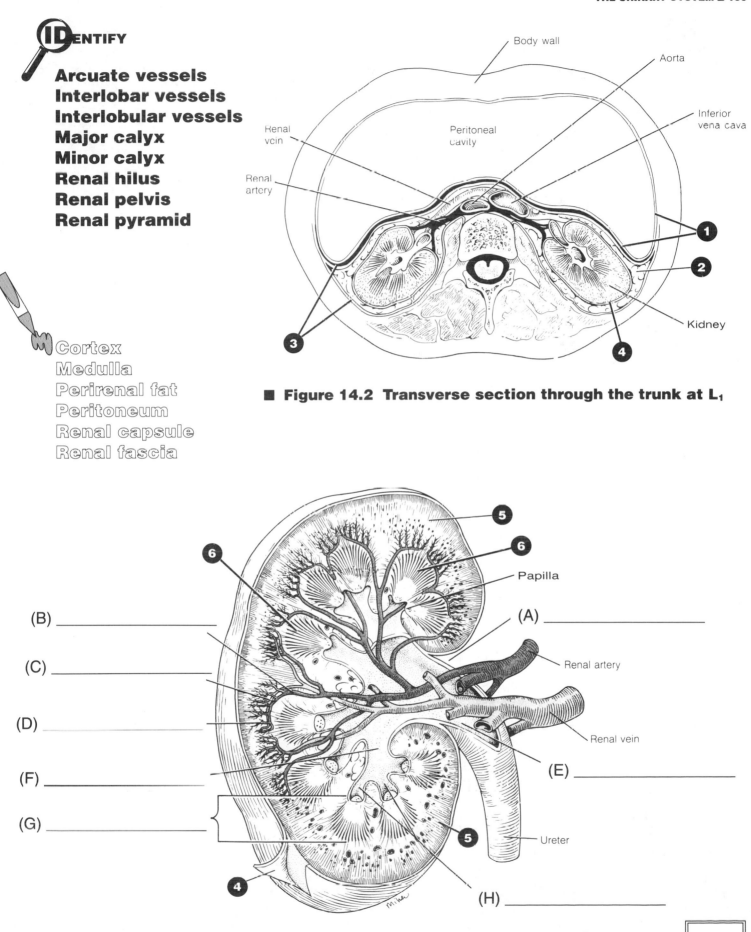

IDENTIFY

Arcuate vessels
Interlobar vessels
Interlobular vessels
Major calyx
Minor calyx
Renal hilus
Renal pelvis
Renal pyramid

Cortex
Medulla
Perirenal fat
Peritoneum
Renal capsule
Renal fascia

Body wall
Aorta
Inferior vena cava
Renal vein
Peritoneal cavity
Renal artery
Kidney

① ② ③ ④

■ **Figure 14.2 Transverse section through the trunk at L₁**

(B) _____

(C) _____

(D) _____

(F) _____

(G) _____

(A) _____

Renal artery

Renal vein

(E) _____

Ureter

(H) _____

Papilla

⑤ ⑥ ④ ⑤ ⑥

■ **Figure 14.3 Longitudinal section through a kidney**

Q1-7

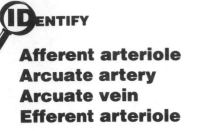

IDENTIFY

Afferent arteriole   **Glomerulus**
Arcuate artery   **Interlobular artery**
Arcuate vein   **Interlobular vein**
Efferent arteriole   **Peritubular capillaries**

Bowman's capsule   Loop of Henle: ascending limb
Collecting duct   Loop of Henle: descending limb
Distal convoluted tubule Proximal convoluted tubule

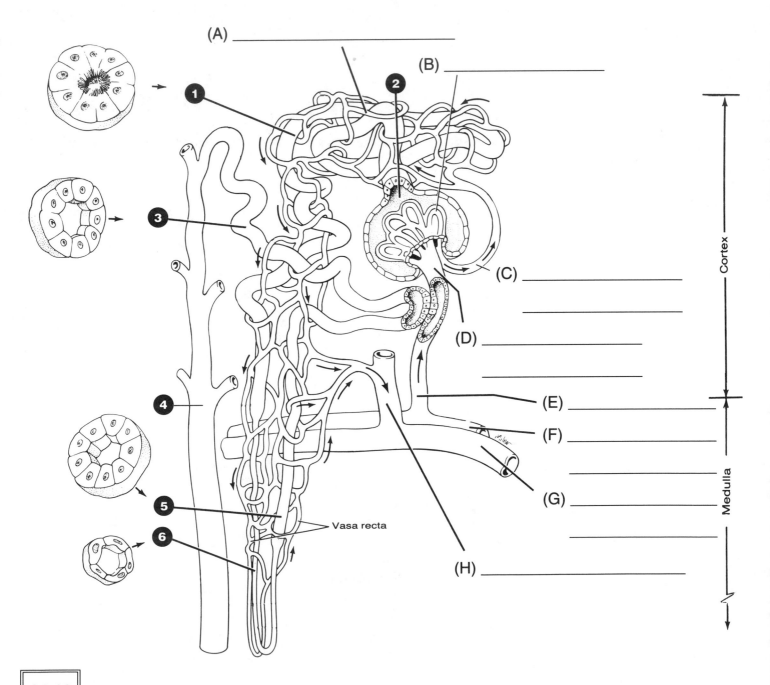

(A) _____

(B) _____

(C) _____

(D) _____

_____

(E) _____

(F) _____

_____

(G) _____

(H) _____

Cortex

Medulla

Vasa recta

Q8-16

■ Figure 14.4 Detail of a nephron and its blood supply (400x)

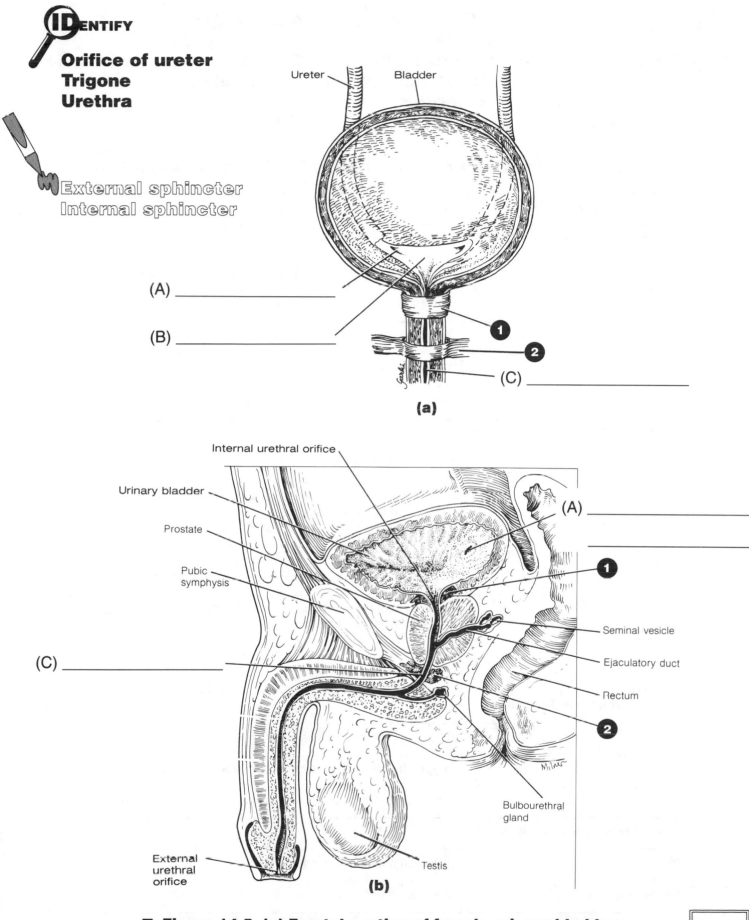

IDENTIFY

Orifice of ureter
Trigone
Urethra

External sphincter
Internal sphincter

Ureter

Bladder

(A) _____

(B) _____

1

2

(C) _____

(a)

Internal urethral orifice

Urinary bladder

Prostate

Pubic
symphysis

(A) _____
_____

1

Seminal vesicle

Ejaculatory duct

Rectum

(C) _____

2

Bulbourethral
gland

External
urethral
orifice

Testis

(b)

■ **Figure 14.5 (a) Frontal section of female urinary bladder,
and (b) sagittal section of male pelvis and urinary bladder**

Q17-20

IDENTIFY

	Distal convoluted tubule	Major calyces
Afferent arterioles	Interlobar arteries	Minor calyces
Arcuate arteries	Interlobular arteries	Proximal convoluted tubule
Bowman's capsule	Glomerulus	Renal pelvis
Collecting duct	Loop of Henle	Ureter
		Urethra

■ **Table 14.1 Summary of Excretory Pathway Through the Urinary System**

Component	Description
Incoming Blood Vessels	
A.	Branches of renal artery travel between the pyramids.
B.	Curved segments at the medullary/cortical junction.
C.	Branches extend through cortex toward kidney surface.
D.	Arterioles supply renal corpuscles.
E.	Capillary network within renal corpuscle.
Tubular Structures of Nephron	
F.	Double walled cup of squamous epithelium.
G.	Twisted tubule in cortex is connected to renal capsule. Composed of simple cuboidal epithelium.
H.	Hairpin segment dips into medulla and returns to cortex. Descending segment is composed of squamous epithelium; ascending loop is composed of cuboidal epithelium.
I.	Twisted tubule in cortex near the end of the nephron. Composed of simple cuboidal epithelium.
J.	Tubule accepts urine from numerous nephrons. Extends through and forms renal pyramids.
Urine Collecting Structures	
K.	Small cup like chambers accept urine from renal papillae.
L.	Larger chambers formed by two or three smaller cups.
M.	Single, large basin or collecting area protrudes from kidney.
Transfer/Storage Structures	
N.	Retroperitoneal tube conveys urine to the urinary bladder.
Urinary Bladder	Muscular reservoir for urine.
O.	Muscular tube carries urine to the outside of the body.

Q21

MODULE 15

The Reproductive System

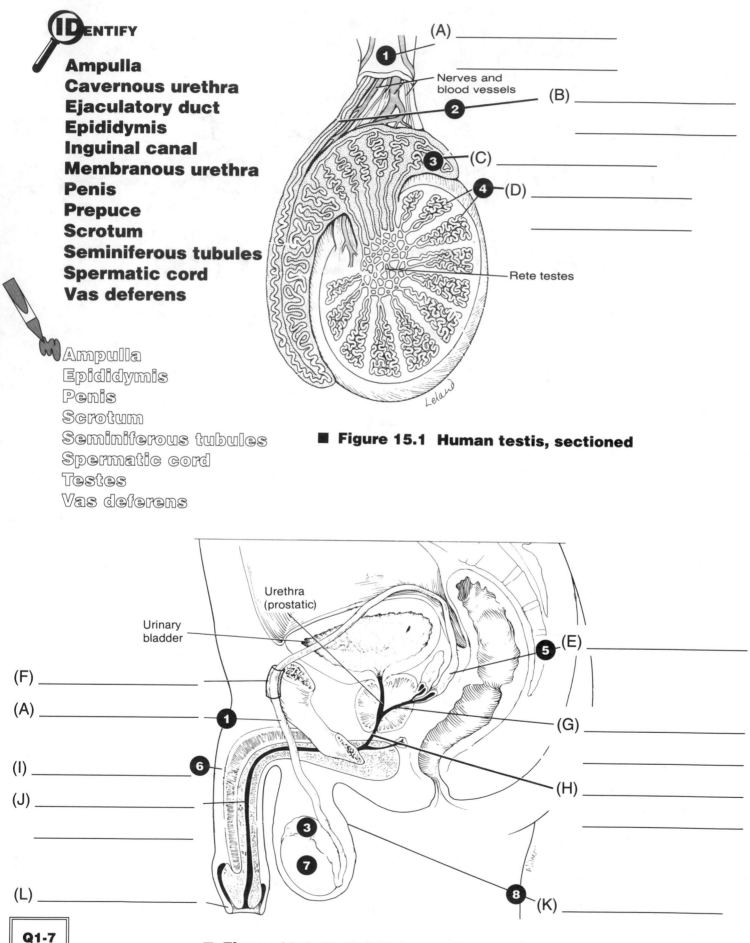

IDENTIFY

Ampulla
Cavernous urethra
Ejaculatory duct
Epididymis
Inguinal canal
Membranous urethra
Penis
Prepuce
Scrotum
Seminiferous tubules
Spermatic cord
Vas deferens

Ampulla
Epididymis
Penis
Scrotum
Seminiferous tubules
Spermatic cord
Testes
Vas deferens

(A) _____

1

Nerves and
blood vessels

(B) _____

2

_____

(C) _____

3

(D) _____

4

_____

Rete testes

Leland

■ **Figure 15.1 Human testis, sectioned**

Urethra
(prostatic)

Urinary
bladder

(E) _____

5

(F) _____

(G) _____

(A) _____

1

(H) _____

(I) _____

6

(J) _____

_____

3

7

(L) _____

8

(K) _____

Q1-7

■ **Figure 15.2 Male pelvis, sagittal section**

IDENTIFY

Bulb
Bulbourethral gland
Corpora cavernosa
Corpus spongiosum
Crus   (A) _____
Glans
Prostate   _____
Seminal vesicle

(B) _____

(C) _____

_____

(E) _____

(F) _____

(D) _____

_____

(G) _____

_____

(H) _____

Ureters

Urinary bladder

Symphysis pubis

(a)

Ampulla
Bulb
Bulbourethral gland
Corpus cavernosum
Corpus spongiosum
Crus
Epididymis
Glans
Prostate
Seminal vesicle
Testes
Vas deferens

(b)

Dorsal nerves and vessels

Central artery of penis

Urethra

(c)

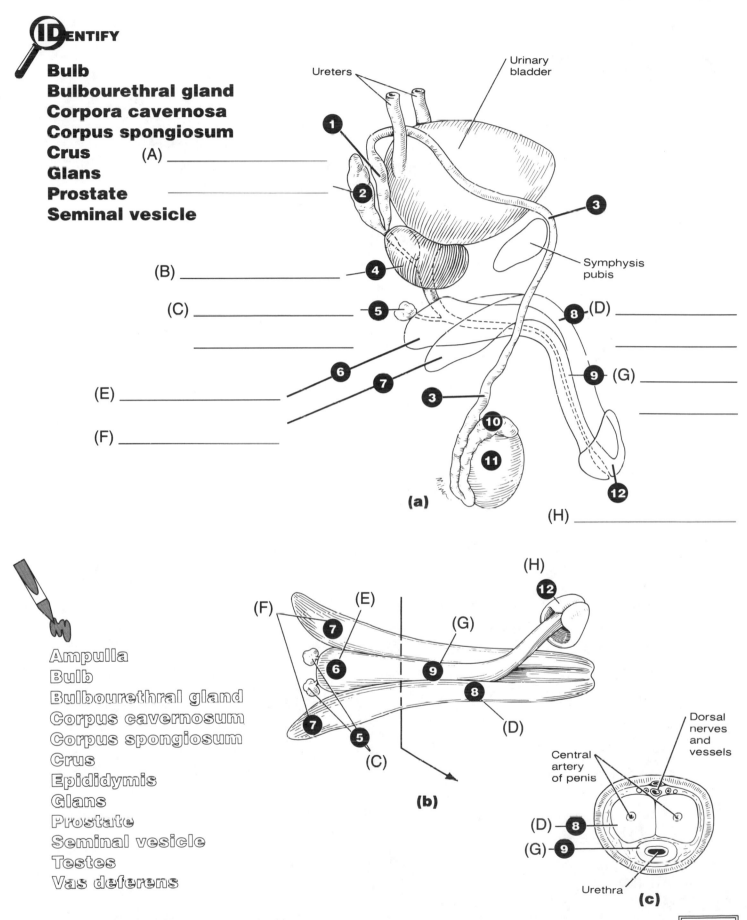

■ Figure 15.3 Male genitalia: (a) tubes and accessory glands,
(b) penis, dissected, inferior view, and (c) penis, cross section

Q8-12

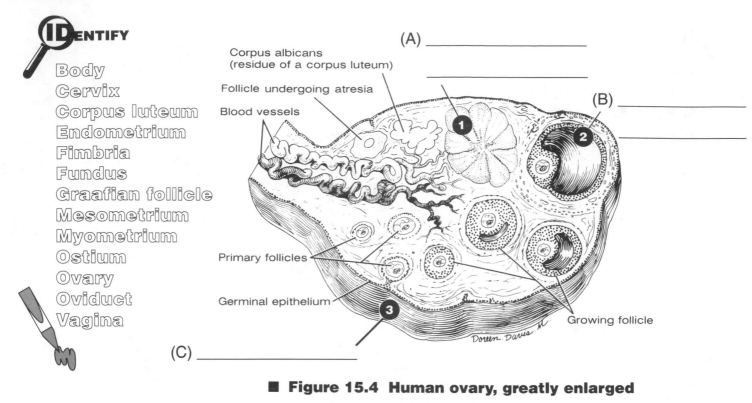

IDENTIFY

Body
Cervix
Corpus luteum
Endometrium
Fimbria
Fundus
Graafian follicle
Mesometrium
Myometrium
Ostium
Ovary
Oviduct
Vagina

Corpus albicans
(residue of a corpus luteum)

Follicle undergoing atresia

Blood vessels

Primary follicles

Germinal epithelium

Growing follicle

(A) _____

_____

(B) _____

_____

(C) _____

■ **Figure 15.4 Human ovary, greatly enlarged**

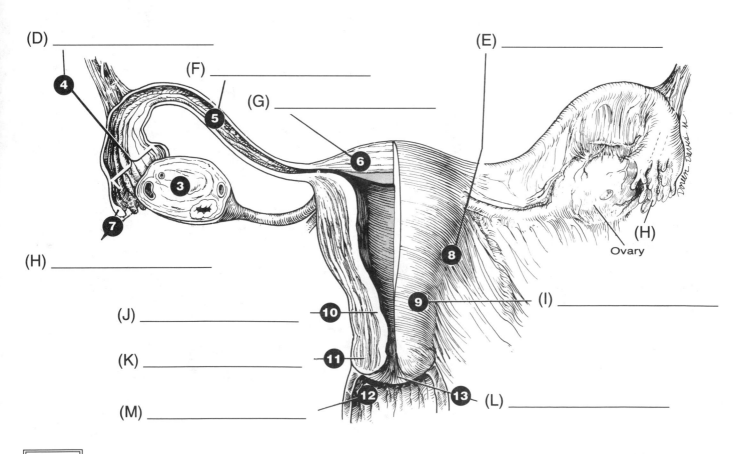

(D) _____

(F) _____

(G) _____

(E) _____

(H) _____

Ovary

(H) _____

(I) _____

(J) _____

(K) _____

(L) _____

(M) _____

■ **Figure 15.5 Female reproductive structures, posterior view**

IDENTIFY

Broad ligament
Clitoris
Hymen
Labia majora
Labia minora
Mesometrium
Mesosalpinx
Mesovarium
Mons pubis
Ovarian ligament
Perineum
Prepuce
Round ligament
Suspensory ligament
Vaginal orifice
Vestibule

Labia majora
Labia minora
Mesometrium
Mesosalpinx
Mesovarium
Mons pubis
Ovarian ligament
Perineum
Prepuce
Round ligament
Suspensory ligament

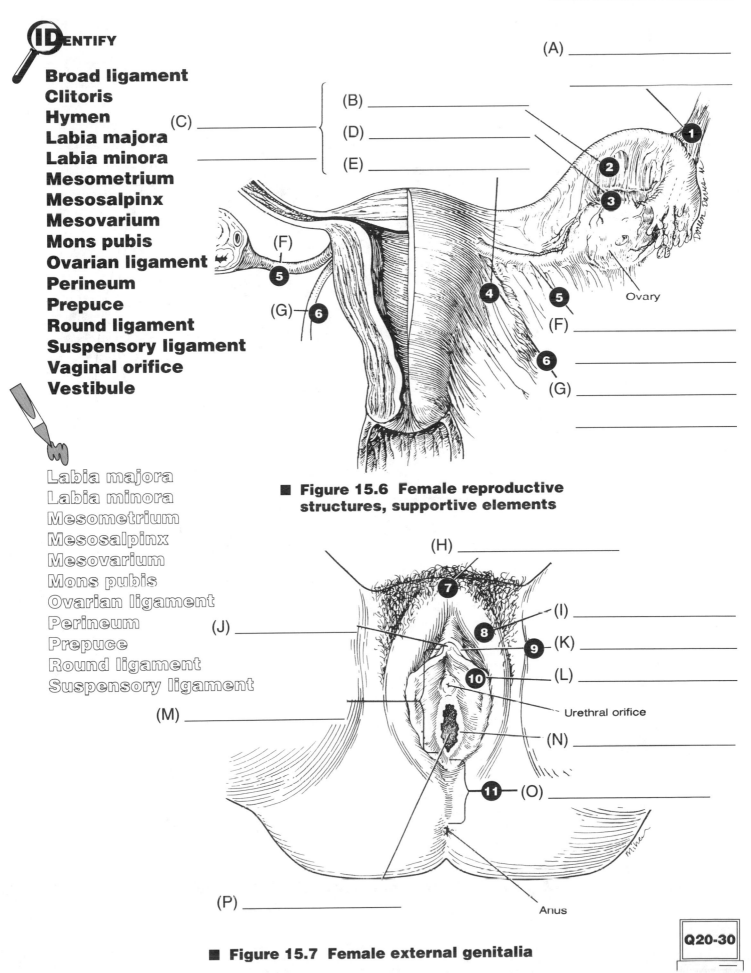

(A) _____

(B) _____

(C) _____

(D) _____

(E) _____

(F) _____

(F) _____

(G) _____

Ovary

(G) _____

_____

■ **Figure 15.6 Female reproductive
structures, supportive elements**

(H) _____

(J) _____

(I) _____

(K) _____

(L) _____

Urethral orifice

(M) _____

(N) _____

(O) _____

(P) _____

Anus

Q20-30

■ **Figure 15.7 Female external genitalia**

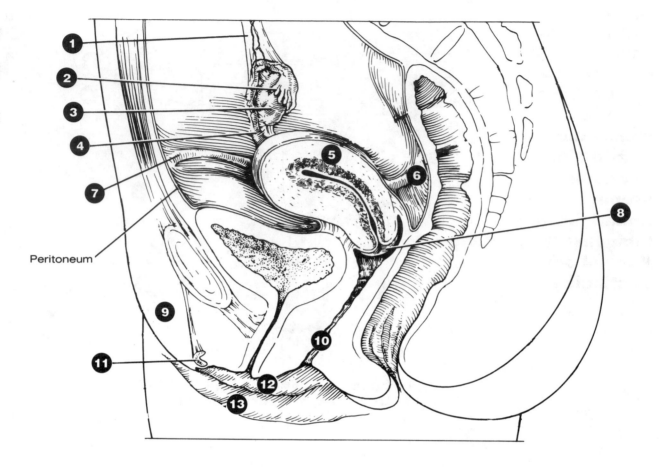

Peritoneum

Broad ligament   Labia minora   Round ligament
Cervix   Mons pubis   Suspensory ligament
Clitoris   Ovary   Uterus
Fimbria   Oviduct   Vagina
Labia majora

■ Figure 15.8 Female pelvis, sagittal section

Q23-30

ANSWER KEY

MODULE 1:
ANATOMICAL TERMINOLOGY

Page 2
Table 1.1 Common Anatomical Terminology

Term	Definition or Common Meaning
Anterior (ventral)	The front
Posterior (dorsal)	The back
Superior (cranial)	Higher
Inferior (caudal)	Lower
Medial	Toward the center or midline
Lateral	Away from the center or midline
Proximal	Toward the origin or beginning of a structure
Distal	Toward the end of a structure
Prone	Lying horizontally with face down
Supine	Lying horizontally with face up
Superficial	Located near or at the surface
Deep	Located away from a surface
Dorsum	Top portion of the foot
Plantar	Sole of the Foot
Palmar	Anterior hand; the palm
Groin (inguinal)	Junction of thigh with abdomen
Axilla	Inferior surface at junction of arm and body; the armpit

Page 3
Figure 1.1 Directional and regional terms illustrated, (a) lateral and (b) anterior persepectives

Identify
A. Axilla
B. Medial
C. Lateral
D. Superior
E. Palmar
F. Proximal
G. Groin
H. Dorsum
I. Distal
J. Inferior

Definition
Standard Anatomical Position:
Body erect, feet together
Hands at side, palms facing foward (anterior)
Fingers extended, thumbs facing away from body

Page 4
Figure 1.2 Body planes

Color
1. Sagittal
2. Frontal
3. Transverse

Definition
a. Longitudinal; divides into anterior/posterior
b. Longitudinal; divides into left/right
c. Horizontal; divides into superior/inferior

Page 5
Figure 1.3 Major body cavities, shown in mid-sagittal section

Color
1. Cranial cavity
2. Spinal cavity
3. Thoracic cavity
4. Abdominal cavity
5. Pelvic cavity

List
a. Brain
b. Spinal cord
c. Heart and pericardium, lungs and pleura
d. Digestive organs
e. Bladder, reproductive organs

Page 6
Figure 1.4 Abdominal regions

Color
1. Right hypochondriac
2. Epigastric
3. Left hypochondriac
4. Right lumbar
5. Umbilical
6. Left lumbar
7. Right iliac
8. Hypogastric
9. Left iliac

MODULE 2: HISTOLOGY

Page 9
Figure 2.1 Simple squamous epithelium

Color
1. Cytoplasm of squamous cells

Figure 2.2 Simple cuboidal epithelium

Color
1. Cytoplasm of cuboidal cells
2. Nuclei of cuboidal cells
3. Connective tissue deep to epithelium

Figure 2.3 Simple columnar epithelium

Color
1. Cytoplasm of columnar cells
2. Nuclei of columnar cells
3. Connective tissue deep to epithelium

Figure 2.4 Pseudostratified ciliated columnar epithelium

Color
1. Cytoplasm of pseudostratified cells
2. Nuclei of pseudostratified columnar cells
3. Connective tissue deep to epithelium
4. Cilia of pseudostratified columnar cells

Page 10
Figure 2.5 Stratified squamous epithelium (mucous membrane)

Color
1. Cytoplasm of squamous cells
2. Nuclei of squamous cells
3. Connective tissue deep to epithelium

Figure 2.6 Stratified columnar epithelium

Color
1. Cytoplasm of columnar cells
2. Nuclei of columnar cells
3. Connective tissue deep to epithelium

Figure 2.7 Transitional epithelium

Color
1. Cytoplasm of transitional cells
2. Nuclei of deep cells
3. Connective tissue deep to epithelium

Figure 2.8 Stratified squamous epithelium (skin)

Color
1. Cytoplasm of squamous cells
2. Nuclei of squamous cells
3. Connective tissue deep to epithelium

Page 11
Figure 2.9 Simple cuboidal epithelium
Figure 2.10 Stratified columnar epithelium
Figure 2.11 Simple squamous epithelium
Figure 2.12 Transitional epithelium
Figure 2.13 Simple columnar epithelium
Figure 2.14 Pseudostratified ciliated columnar epithelium
Figure 2.15 Stratified squamous epithelium

Identify
A. Cuboidal cells
B. Columnar cells
C. Squamous cells
D. Transitional cells
E. Pseudostratified columnar cells

Page 12
Figure 2.16 *Areolar connective tissue*

Color
1. Collagenous fibers
2. Ground substance

Figure 2.17 *Dense irregular connective tissue*

Color
1. Collagenous fibers
2. Ground substance

Figure 2.18 *Dense regular connective tissue*

Color
1. Collagenous fibers

Figure 2.19 *Elastic connective tissue*

Color
1. Collagenous fibers

Page 13
Figure 2.20 *Adipose tissue*

Color
1. Cytoplasm

Figure 2.21 *Hyaline cartilage*

Color
1. Cytoplasm of chondrocyte
2. Matrix of hyaline cartilage

Figure 2.22 *Elastic cartilage*

Color
1. Cytoplasm of chondrocyte
2. Matrix of elastic cartilage

Figure 2.23 *Fibrocartilage*

Color
1. Cytoplasm of chondrocyte
2. Matrix of fibrocartilage
3. Collagenous fibers

Page 14
Figure 2.24 *Bone*
Figure 2.25 *Hyaline cartilage*

Color
1. Nuclei of chondrocytes
2. Cytoplasm of chondrocytes
3. Matrix of hyaline cartilage

Figure 2.26 *Dense irregular connective tissue*

Color
1. Collagenous fibers
2. Ground substance

Figure 2.27 *Elastic tissue*

Color
1. Collagenous fibers
2. Ground substance

Page 15
Figure 2.28 *Elastic cartilage*

Color
1. Matrix of elastic cartilage
2. Cytoplasm of chondrocytes

Figure 2.29 *Areolar connective tissue*

Color
1. Collagenous fibers
2. Nuclei of fibroblasts
3. Ground substance

Figure 2.30 *Fibrocartilage*

Color
1. Collagenous fibers
2. Chrondrocyte nuclei

Figure 2.31 *Dense regular connective tissue*

Color
1. Collagenous fibers
2. Nuclei of fibroblasts

Page 16
Figure 2.32 *Skeletal muscle (long section, featuring unbranched fibers)*

Color
1. Cytoplasm (sarcoplasm) of skeletal muscle cell
2. Nuclei (small, numerous, peripherally placed)

Figure 2.33 *Cardiac muscle (long section, featuring branched fibers)*

Color
1. Cytoplasm of cardiac muscle cell
2. Nucleus (central, large, single)

Figure 2.34 *Smooth muscle (long section, featuring spindle-shaped fibers)*

Color
1. Cytoplasm of smooth muscle cell
2. Nuclei (central, single)

Figure 2.35 *Smooth muscle, long section and cross section*

Color
1. Cytoplasm of smooth muscle cells
2. Nuclei of smooth muscle cells

Page 17
Figure 2.36 *Cardiac muscle, cross section*

Color
1. Cytoplasm
2. Nuclei (centrally placed)

Figure 2.37 *Skeletal muscle, cross section*

Color
1. Cytoplasm
2. Nuclei (peripherally placed)

Figure 2.38 *Nervous tissue in the spinal cord (cross section)*

Color
1. Myelinated fibers (white matter)
2. Nuclei of multipolar neurons

Figure 2.39 *Multipolar neurons*

Color
1. Cytoplasm of multipolar neuron
2. Nuclei of multipolar neurons
3. Dendrites
4. Axon

MODULE 3: THE SKELETON

Page 20
Table 3.1 Bone Markings

Name	Description and/or Example
Process	Any prominent projection, styloid process
Crest	A sharp, semicircular ridge, iliac crest
Spine	A slender, pointed projection, ischial spine
Epicondyle	A small nodule near a condyle, medial femoral epicondyle
Tubercle	A small, rough nodule, lesser tubercle
Tuberosity	A large, rough nodule, ischial tuberosity
Trochanter	A large, blunt process, greater trochanter
Condyle	A smooth rounded surface, femoral condyles
Facet	A shallow, nearly flat pad, costal articular facet
Fissure	A narrow, cleftlike passage, orbital fissures
Foramen	A hole, foramen magnum
Meatus	A tubular canal, external auditory meatus
Sinus	A concealed, hollow cavity, paranasal sinuses
Sulcus	A groove, transverse sulcus

Page 21
Figure 3.1 Divisions of the skeleton

Color
1. Axial division
2. Appendicular division

Page 22
Figure 3.2 Transverse section through cranial cavity

Identify
A. Ethmoid
B. Frontal
C. Sphenoid
D. Parietal
E. Temporal
F. Occipital

Color

Cranial bone	Count
1. Ethmoid	1
2. Frontal	1
3. Sphenoid	1
4. Parietal	2
5. Temporal	2
6. Occipital	1

Page 23
Figure 3.3 Anterior view of the skull

Identify
A. Nasal
B. Lacrimal
C. Zygomatic
D. Inferior nasal conchae
E. Vomer
F. Maxillae
G. Mandible

Color

Facial bones	Count
1. Nasal	2
2. Lacrimal	2
3. Zygomatic	2
4. Inferior nasal conchae	2
5. Maxillae	2
6. Vomer	1
7. Mandible	1

Page 24
Figure 3.4 Right orbital cavity

Identify
A. Superior orbital fissure
B. Inferior orbital fissure

Color
1. Frontal
2. Sphenoid
3. Palatine
4. Ethmoid
5. Lacrimal
6. Zygomatic
7. Maxilla

Page 25
Figure 3.5 Left lateral (internal) view of nasal cavity

Color
1. Frontal
2. Nasal
3. Sphenoid
4. Ethmoid
5. Inferior nasal concha
6. Palatine
7. Maxillae

Visible subparts
a. Superior nasal concha
b. Middle nasal concha
c. Frontal sinus
d. Pterygoid process
e. Sphenoidal sinus

Page 26
Figure 3.6 Sagittal view of the Skull

Color
1. Parietal
2. Frontal
3. Temporal
4. Sphenoid
5. Ethmoid
5a. Perpendicular plate
6. Nasal
7. Occipital
8. Vomer
9. Palatine
10. Maxillae
11. Mandible

Bony components of the nasal septum (any order)
a. Perpendicular plate
b. Vomer

Cranial bones (any order)
c. Ethmoid
d. Frontal
e. Occipital
f. Parietal
g. Sphenoid
h. Temporal

Facial Bones (any order)
I. Mandible
j. Maxillae
k. Nasal

l. Palatine
m. Vomer

Page 27
Figure 3.7 Paranasal sinuses

Color
1. Frontal sinus
2. Ethmoidal sinus
3. Sphenoidal sinus
4. Maxillary sinus

Figure 3.8 Middle ear cavity: auditory ossicles

Color
1. Malleus
2. Incus
3. Stapes

Page 28
Figure 3.9 Lateral view of the skull

Identify
A. Coronal (frontal) suture
B. Lambdoidal suture
C. Squamosal suture

Color
1. Parietal
2. Frontal
3. Sphenoid
4. Ethmoid
5. Lacrimal
6. Nasal
7. Occipital
8. Temporal
9. Zygomatic
10. Maxilla
11. Mandible
12. Hyoid

Page 29
Figure 3.10 Inferior view of the skull

Identity
A. Incisive foramen (fossa)
B. Mastoid process
C. Stylomastoid foramen
D. Jugular foramen (fossa)
E. Occipital condyle

Color
1. Maxilla
2. Inferior nasal conchae
3. Palatine
4. Zygomatic
5. Vomer
6. Sphenoid
7. Temporal
8. Parietal
9. Occipital

Page 30
Figure 3.11 Sphenoid relationships in cranium
Figure 3.12 Sphenoid bone in (a) superior and (b) posterior vews

Identify
A. Optic foramen
B. Foramen rotundum
C. Foramen ovale
D. Sella turcica
E. Internal auditory meatus
F. Jugular foramen
G. Pterygoid processes

Page 31
Figure 3.13 Right temporal bone, lateral view
Figure 3.14 Right maxilla and vicinity, lateral view
Figure 3.15 Left palatine bone, posterior view

Identify
A. External auditory meatus
B. Mastoid proces
C. Styloid process
D. Lacrimal duct or sulcus
E. Sphenopalatine foramen

Page 32
Figure 3.16 Lateral view of vertebral column and superior views of three typical vertebrae

Identify
A. Cervical region
B. Intervertebral disc
C. Thoracic region
D. Intervertebral foramen
E. Lumbar region
F. Sacral region
G. Coccygeal region

Color
1. Cervical region
2. Thoracic region
3. Lumbar region
4. Sacral region
5. Coccygeal region

Vertebral Region (any order)	Count
Cervical	7
Coccygeal	4
Lumbar	5
Sacral	5
Thoracic	12

Page 33
Figure 3.17 Typical vertebra in (a) superior and (b) lateral views

Identify
A. Spinous process
B. Superior articular process
C. Lamina
D. Pedicle
E. Transverse process
F. Inferior articular process
G. Body (centrum)

Figure 3.18 Typical cervical vertebra, superior view

Figure 3.19 First and second cervical vertebrae

Color
1. Spinous process
2. Lamina
3. Pedicle
4. Body or centrum
5. Axis
6. Atlas

Page 34
Figure 3.20 Two typical thoracic vertebrae, posterolateral view

Identify
A. (Articular) facet
B. Demifacet

Color
1. Superior articular processes
2. Body of vertebrae
3. Interior articular processes

Figure 3.21 Articulation of rib and thoracic vertebra, superior view

Identify
A. (Articular) facet
B. Demifacet

Color
1. Superior articular processes
2. Body of vertebrae
3. Inferior articular processes

Page 35
Figure 3.22 Two typical lumbar vertebrae, posterolateral view

Color
1. Superior articular process
2. Body of vertebrae
3. Inferior articular processes
4. Intervertebral disc

Figure 3.23 Sacrum and coccyx in (a) anterior and (b) posterior views

Color
1. Superior articular process
2. Body of vertebrae
3. Inferior articular processes
4. Intervertebral disc

Page 36
Figure 3.24 Thorax, anterior view

Identify
A. Manubrium
B. Body
C. Xiphoid process

Color
1. True ribs
2. Costal cartilage
3. False ribs

Figure 3.25 Right rib, posterior view

Identify
A. Head
B. Neck
C. Tubercle
D. Shaft

Page 37
Figure 3.26 (a) Left shoulder girdle, anterior view (b) left scapula, posterior view (c) left scapula, anterior view

Identify
A. Acromion process
B. Coracoid process
C. Superior border
D. Supraspinous fossa
E. Glenoid fossa
F. Vertebral border
G. Infraspinous fossa
H. Axillary border
I. Subscapular fossa

Color
1. Acromion process
2. Coracoid process
3. Supraspinous fossa
4. Glenoid fossa
5. Infraspinous fossa
6. Subscapular fossa

Page 38
Figure 3.27 Right humerus in (a) anterior and (b) posterior views

Identify
A. Greater tubercle
B. Lesser tubercle
C. Anatomical neck
D. Surgical neck
E. Medial epicondyle
F. Lateral epicondyle
G. Capitulum
H. Trochlea
I. Olecranon fossa

Color
1. Greater tubercle
2. Lesser tubercle
3. Medial epicondyle
4. Capitulum
5. Trochlea
6. Lateral epicondyle

Page 39
Figure 3.28 Right radius and ulna, anterior view

Identify
A. Olecranon process
B. Coronoid process
C. Trochlear (semilunar) notch
D. Styloid process of radius
E. Styloid process of ulna

Color
1. Radial head
2. Radial notch (on ulna)
3. Radial tuberosity

Page 40
Figure 3.29 Bones of the right wrist and hand, anterior view

Identify
A. Fourth middle phalanx
B. Distal
C. Phalanges
D. Middle
E. Proximal
F. Metacarpals
G. Carpals

Color
1. Hamate
2. Capitate
3. Trapezoid
4. Trapezium
5. Pisiform
6. Triquetral
7. Lunate
8. Scaphoid

Proximal row of carpals (lateral to medial)
a. Scaphoid
b. Lunate
c. Triquetral
d. Pisiform

Distal row of carpals (lateral to medial)
e. Trapezium
f. Trapezoid
g. Capitate
h. Hamate

Page 41
Figure 3.30 Anterior views of (a) male pelvis and (b) female pelvis

Identify
A. Sacroiliac joint
B. Ischial spine

Color
1. Sacrum
2. Coccyx
3. Symphysis pubis
4. Acetabulum

Page 42
Figure 3.31 Right coxal bone in (a) internal view

Identify
A. Anterior superior iliac spine
B. Posterior superior iliac spine
C. Anterior inferior iliac spine
D. Posterior inferior iliac spine
E. Greater sciatic notch
F. Ischial spine
G. Obturator foramen

Color
1. Ilium
2. Pubis
3. Ischium

Page 43
Figure 3.31, continued, Right coxal bone in (b) external view

Identify
A. Anterior superior iliac spine
B. Posterior superior iliac spine
C. Anterior inferior iliac spine
D. Posterior inferior iliac spine
E. Greater sciatic notch
F. Ischial spine
G. Obturator foramen
H. Ischial tuberosity

Color
1. Ilium
2. Pubis
3. Ischium

Page 44
Figure 3.32 Right femur in (a) anterior and (b) posterior views

Identify
A. Fovea capitis
B. Greater trochanter
C. Lesser trochanter
D. Linea aspera
E. Adductor tubercle
F. Medial epicondyle
G. Lateral epicondyle
H. Lateral condyle
I. Medial condyle

Color
1. Greater trochanter
2. Lesser trochanter
3. Linea aspera
4. Lateral condyle
5. Medial condyle

Page 45
Figure 3.33 Patella, anterior and posterior views

Color
1. Medial articular surface
2. Lateral articular surface

Figure 3.34 Right tibia and fibula in (a) anterior and (b) posterior views

Identify
A. Lateral condyle
B. Medial condyle
C. Tibial tuberosity

D. Medial malleolus (on tibia)
E. Lateral malleolus (on fibula)

Page 46
Figure 3.35 Right ankle and foot, superior view

Identify
A. First proximal phalanx
B. Distal
C. Phalanges
D. Middle
E. Proximal
F. Metatarsals
G. Tarsals

Color
1. Medial cuneiform
2. Intermediate cuneiform
3. Lateral cuneiform
4. Cuboid
5. Navicular
6. Talus
7. Calcaneus

Figure 3.36 Right ankle and foot in (a) medial and (b) lateral views

Color
1. Medial cuneiform
2. Intermediate cuneiform
3. Lateral cuneiform
4. Cuboid
5. Navicular
6. Talus
7. Calcaneus

MODULE 4: ARTHROLOGY

Page 48
Figure 4.1 Functional and structural joint designations

Identify
A. Synarthrosis
B. Suture
C. Amphiarthrosis
D. Synchondrosis
E. Amphiarthrosis
F. Symphysis
G. Amphiarthrosis
H. Symphysis
I. Synarthrosis
J. Syndesmosis
K. Diarthrosis
L. Synovium
M. Diarthrosis
N. Synovium
O. Diarthrosis
P. Synovium

Color
1. Bone
2. Fibrous connective tissue
3. Hyaline cartilage
4. Fibrocartilage
5. Articular capsule
6. Articular cartilage

Page 50
Table 4.1 Summary of Human Joint Classification

Identify
A. Synarthroses
B. Sutures
C. Syndesmoses
D. Amphiarthroses

E. Symphyses
F. Synchondroses
G. Diarthroses
H. Synovia

Examples
a. Squamosal suture (Figure 4.1a)
b. Distal tibia/fibula (Figure 4.1e)
c. Between vertebral bodies (Figure 4.1c)
d. Pubic symphysis (Figure 4.1d)
e. Rib/sternum via costal cartilages (Figure 4.1b)
f. Gliding: intercarpal articulations (Figure 4.1h)
g. Humerus/ulna (hinge), radius/ulna (pivot) (Figure 4.1g)
h. Condyloid: Radius/carpals (Figure 4.1h)
i. Scapula/humerus (Figure 4.1f)

Page 51
Figure 4.2 Axes of rotation of diarthrotic joints

Identify
A. Uniaxial
B. Biaxial
C. Triaxial

Page 52
Figure 4.3 Diarthrotic joint features

Identify
A. Synovial membrane
B. Synovial cavity with synovial fluid
C. Meniscus (articular disc)
D. Articular cartilage
E. Articular capsule

Color
1. Synovial membrane
2. Synovial cavity with synovial fluid
3. Meniscus (articular disc)
4. Articular cartilage
5. Articular capsule

Page 53
Figure 4.4 Ligaments of the knee
Figure 4.5 Knee, medial view with femur sectioned

Identify
A. Lateral collateral ligament
B. Posterior cruciate ligament
C. Anterior cruciate ligament
D. Lateral meniscus
E. Medial meniscus
F. Medial collateral ligament

Color
1. Lateral collateral ligament
2. Posterior cruciate ligament
3. Anterior cruciate ligament
4. Lateral meniscus
5. Medial meniscus
6. Medial collateral ligament

Page 54
Figure 4.6 Modified synovial membranes

Color
1. Bone
2. Skin
3. Tendon sheath
4. Tendon
5. Synovial cavity
6. Synovial membrane

MODULE 5: MUSCLE ACTIONS

Page 56
Figure 5.1 *Points of muscle attachment*

Identify
A. Origin
B. Insertion

Definition
a. Immovable point of attachment
b. Movable point of attachment

Figure 5.2 *Flexion*
Figure 5.3 *Extension*
Figure 5.4 *Abduction*
Figure 5.5 *Adduction*

Page 57
Figure 5.6 *Rotation*
Figure 5.7 *Circumduction*
Figure 5.8 *Pronation*
Figure 5.9 *Supination*
Figure 5.10 *Protraction*
Figure 5.11 *Retraction*

Page 58
Figure 5.12 *Elevation*
Figure 5.13 *Depression*
Figure 5.14 *Inversion*
Figure 5.15 *Eversion*
Figure 5.16 *Plantar flexion*
Figure 5.17 *Dorsiflexion*

MODULE 6: SKELETAL MUSCLES

No.	Muscle	Pg.	Origin	Insertion	Action
			Muscles of the Face		
1	Orbicularis oculi	64	Frontal and maxillary bones	Skin around orbit	Depresses eyelids
2	Orbicularis oris	64	Muscles around mouth	Skin around mouth	Protrudes (purses) lips
3	Buccinator	64	Maxilla and mandible	Angle of mouth	Retracts angle of mouth
4	Platysma	64	Fascia of deltoid muscle	Mandible and skin around lower lip	Depresses mandible and lower lip
			Muscles of Mastication		
5	Temporalis	66	Temporal fossa	Coronoid process of mandible	Elevates mandible
6	Masseter	66	Zygomatic arch of temporal	Mandible	Elevates mandible
			Muscles of the Anterior Triangle of the Neck		
7	Stylohyoid	67	Styloid process	Hyoid bone	Elevates hyoid and tongue
8	Thyrohyoid	67	Thyroid cartilage	Hyoid bone	Depresses hyoid bone
9	Mylohyoid	68	Inferior surface of mandible	Hyoid bone	Elevates hyoid and tongue
10	Digastric	68	Mastoid process and inner mandible	Hyoid bone	Depresses mandible/elevates hyoid
11	Sternocleidomastoid	68	Manubrium and clavicle	Mastoid process	Rotates or flexes head
12	Sternohyoid	68	Manubrium	Hyoid bone	Depresses hyoid
13	Sternothyroid	68	Manubrium	Thyroid cartilage	Depresses thyroid cartilage
			Muscles That Act on the Scapula		
14	(Clavo)trapezius	70	Ligamentum nuchae	Clavicle	Flexes head laterally/elevates clavicle
15	Levator scapulae	70	Cervical vertebrae	Vertebral border of scapula	Elevates scapula
16	(Acromio)trapezius	70	Ligamentum nuchae	Acromion process of scapula	Adducts scapula
17	Rhomboideus	70	Spines of thoracic vertebrae	Vertebral border of scapula	Adducts scapula
18	(Spino)trapezius	70	Ligamentum nuchae	Spine of scapula	Depresses scapula
19	Pectoralis minor	72	Upper ribs	Coracoid process of scapula	Depresses scapula
20	Serratus anterior	72	Upper ribs	Vertebral border of scapula	Anterior rotation of scapula
			Muscles That Act on the Arm (Humerus)		
21	Coracobrachialis	72	Coracoid process of scapula	Shaft of humerus	Adducts humerus
22	Subscapularis	72	Subscapular fossa	Lesser tubercle of humerus	Medial rotation of humerus
23	(Acromio)deltoid	74	Acromion process of scapula	Deltoid tuberosity of humerus	Abducts humerus
24	(Clavo)deltoid	74	Clavicle	Deltoid tuberosity of humerus	Flexes humerus
25	Pectoralis major	74	Sternum and clavicle	Greater tubercle of humerus	Flexes and adducts humerus
26	Supraspinatus	76	Supraspinous fossa	Greater tubercle of humerus	Abducts humerus
27	(Spino)deltoid	76	Spine of scapula	Deltoid tuberosity of humerus	Extends humerus
28	Infraspinatus	76	Infraspinous fossa	Greater tubercle of humerus	Lateral rotation of humerus
29	Teres major	76	Lateral border of scapula	Lesser tubercle of humerus	Adducts humerus
30	Latissimus dorsi	76	Lumbodorsal fascia	Lesser tubercle of humerus	Adduction (extension) of humerus
			Muscles That Act on the Forearm (Radius and Ulna)		
31	Biceps brachii	78	Coracoid process and glenoid rim	Radial tuberosity	Flexes forearm
32	Brachialis	78	Shaft of humerus	Coronoid process of ulna	Flexes forearm
33	Triceps brachii	79	Scapula and shaft of humerus	Olecranon of ulna	Extends forearm
34	Pronator teres	80	Medial epicondyle of humerus and ulnar coronoid	Shaft of radius	Pronation
35	Brachioradialis	80	Supracordylar ridge of humerus	Styloid of radius	Flexes forearm
			Muscles That Act on the Wrist, Hand, and Fingers		
36	Flexor carpi radialis	80	Medial epicondyle of humerus	2nd and 3rd metacarpals	Flexes wrist
37	Palmaris longus	80	Medial epicondyle of humerus	Palmar aponeurosis	Flexes wrist
38	Flexor carpi ulnaris	80	Medial epicondyle of humerus and upper ulna	Medial carpals; 5th metacarpal	Flexes wrist
39	Extensor carpi radialis longus	82	Lateral epicondyle of humerus	2nd metacarpal	Extends wrist
40	Flexor digitorum superficialis	82	Medial epicondyle of humerus and parts of ulna and radius	Middle phalanges (2–5)	Flexes middle phalanges
41	Flexor digitorum profundus	84	Shaft of ulna	Distal phalanges (2–5)	Flexes fingertips
42	Flexor pollicis longus	84	Shaft of radius	First distal phalanx	Flexes thumb
43	Extensor digiti minimi	86	Lateral epicondyle of humerus	Fifth digit	Extends fifth digit
44	Extensor carpi radialis brevis	86	Lateral epicondyle of humerus	3rd metacarpal	Extends wrist
45	Extensor digitorum	86	Lateral epicondyle of humerus	Digits 2–5	Extends fingers
46	Extensor carpi ulnaris	86	Lateral epicondyle of humerus	5th metacarpal	Extends wrist
47	Supinator	88	Lateral epicondyle of humerus and proximal ulna	Shaft of radius	Supination
48	Abductor pollicis longus	88	Ulna and radius	First metacarpal	Abducts thumb
49	Extensor pollicis longus	88	Posterior ulna	First distal phalanx	Extends thumb
50	Abductor pollicis brevis	90	Carpals and flexor retinaculum	Proximal phalanx of thumb	Abducts thumb
51	Adductor pollicis	90	Metacarpals 2 and 3	Proximal phalanx of thumb	Adducts thumb

No.	Muscle	Pg.	Origin	Insertion	Action
			Respiratory Muscles		
52	External intercostals	91	Outer surface of superior rib	Inferior rib	Elevates ribs
53	Internal intercostals	91	Inner surface of inferior rib and costal cartilage	Superior rib and costal cartilage	Depresses ribs
			Muscles That Move the Vertebral Column		
54	Erector spinae group	92	Iliac crest or ribs and vertebrae inferior to insertion	Ribs and vertebrae superior to origin and mastoid process of temporal	Extends head and vertebral column
			Muscles of the Abdominal Wall		
55	External abdominal oblique	94	Lower eight ribs	Linea alba; iliac crest	Compresses abdomen; flexes and rotates vertebral column
56	Rectus abdominis	94	Pubis	Costal cartilages and xiphoid process	Flexes vertebral column; compresses abdomen
57	Transversus abdominis	94	Iliac crest and costal cartilages	Linea alba	Compresses abdomen
58	Internal abdominal oblique	94	Iliac crest; lumbodorsal fascia	Linea alba and lower ribs	Compresses abdomen; flexes and rotates vertebral column
			Muscles That Act on the Thigh (Femur)		
59	Psoas major	96	Vertebral bodies of T_{12}-L_5	Lesser trochanter of femur	Flexes thigh
60	Iliacus	96	Iliac crest and fossa	Lesser trochanter of femur	Flexes thigh
61	Pectineus	96	Pubis	Proximal femur	Adducts thigh
62	Gluteus medius	98	Iliac crest	Greater trochanter of femur	Abducts and medially rotates thigh
63	Gluteus maximus	98	Iliac crest and sacrum	Fascia lata and proximal femur	Extends and laterally rotates thigh
64	Tensor fasciae latae	100	Iliac crest	Tibia via fascia lata	Abducts thigh; tenses fascia latae
65	Adductor brevis	100	Pubis	Linea aspera of femur	Adducts and laterally rotates thigh
66	Adductor longus	100	Pubis	Linea aspera of femur	Adducts, laterally rotates, and flexes thigh
67	Adductor magnus	100	Pubis and ischium	Linea aspera of femur	Adducts, extends, and laterally rotates thigh
			Muscles That Act on the Leg (Muscles of the Thigh)		
68	Sartorius	102	Anterior superior iliac spine	Proximal medial tibia	Flexes and laterally rotates thigh
69	Gracilis	102	Pubis	Medial tibia	Adducts thigh
70	Rectus femoris	102	Anterior inferior iliac spine	Tibial tuberosity via patella and patellar ligament	Flexes thigh and extends leg
71	Vastus lateralis	102	Greater trochanter and linea aspera of femur	Tibial tuberosity via patella and patellar ligament	Extends leg
72	Vastus medialis	102	Linea aspera of femur	Tibial tuberosity via patella and patellar ligament	Extends leg
73	Vastus intermedius	105	Anterior surface of femur	Tibial tuberosity via patella and patellar ligament	Extends leg
74	Semitendinosus	106	Ischial tuberosity	Medial tibia	Flexes leg, extends thigh
75	Biceps femoris	106	Ischial tuberosity and linea aspera of femur	Tibia at lateral condyle and fibula	Flexes leg, extends thigh
76	Semimembranosus	106	Ischial tuberosity	Tibia at medial condyle	Flexes leg, extends thigh
			Muscles That Act on the Foot and Toes (Muscles of the Leg)		
77	Tibialis anterior	108	Proximal tibia	Medial tarsal and first metatarsal	Dorsiflexes and inverts foot
78	Extensor digitorum longus	108	Proximal tibia and fibula	Phalanges of digits 2–5	Extends digits 2–5; dorsiflexes foot
79	Peroneus longus	110	Proximal fibula	First metatarsal on plantar surface	Plantar flexes and everts foot
80	Peroneus brevis	110	Shaft of fibula	Fifth metatarsal on lateral side	Plantar flexes and everts foot
81	Peroneus tertius	112	Distal fibula	Fifth metatarsal on dorsal surface	Dorsiflexes and everts foot
82	Extensor hallucis longus	112	Middle fibula	Great toe, distal phalanx	Extends great toe
83	Gastrocnemius	114	Femur—medial and lateral condyles	Calcaneus via Achilles (calcaneus) tendon	Plantar flexes foot and flexes leg
84	Soleus	114	Head of fibula and proximal tibia	Calcaneus via Achilles (calcaneus) tendon	Plantar flexes foot
85	Tibialis posterior	116	Posterior tibia and fibula	Tarsals, inferior surface	Plantar flexes and inverts foot
86	Flexor hallucis longus	116	Distal shaft of fibula	Hallux, distal phalanx	Flexes great toe
87	Flexor digitorum longus	118	Posterior shaft of tibia	Distal phalanges of digits 2–5	Flexes digits 2–5
88	Flexor digitorum brevis	119	Calcaneus	Middle phalanges, digits 2–5	Flexes digits 2–5

MODULE 7:
THE CIRCULATORY SYSTEM

Page 122
Figure 7.1 Heart wall, section
Figure 7.2 Human heart, frontal section

Identify
A. Visceral pericardium
B. Parietal pericardium
C. Pulmonary semilunar valve
D. Bicuspid valve
E. Coronary sinus opening
F. Aortic semilunar valve
G. Tricuspid valve
H. Chordae tendineae
I. Interventricular septum
J. Papillary muscle
K. Trabeculae carnae

Color
1. Endocardium
2. Myocardium
3. Aortic arch
4. Superior vena cava
5. Pulmonary trunk
6. Left atrium
7. Right atrium
8. Right ventricle
9. Left ventricle
10. Inferior vena cava
11. Descending aorta

Page 123
Figure 7.3 Aortic semilunar valve
Figure 7.4 Vessels of the heart,
anterior view

Identify
A. Right coronary artery opening
B. Left coronary artery opening
C. Left common carotid artery
D. Left subclavian artery
E. Brachiocephalic artery
F. Aortic arch
G. Superior vena cava
H. Right pulmonary artery
I. Left pulmonary artery
J. Pulmonary trunk
K. Circumflex artery
L. Anterior interventricular artery

Color
1. Vessels carrying deoxygenated blood
2. Vessels carrying oxygenated blood
3. Left atrium
4. Right atrium
5. Right ventricle
6. Left ventricle

Page 124
Figure 7.5 Vessels of the heart,
posterior view

Identify
A. Left subclavian artery
B. Left common carotid artery
C. Aortic arch
D. Brachiocephalic artery
E. Superior vena cava
F. Ligamentum arteriosum
G. Right pulmonary artery
H. Left pulmonary artery
I. Right pulmonary veins
J. Left pulmonary veins
K. Inferior vena cava
L. Coronary sinus
M. Circumflex artery
N. Right coronary artery

Color
1. Vessels carrying oxygenated blood
2. Vessels carrying deoxygenated blood
3. Left atrium
4. Right atrium
5. Left ventricle
6. Right ventricle

Page 125
Figure 7.6 Fetal circulation through the heart
Figure 7.7 Branches of the brachiocephalic
artery

Identify
A. Ductus arteriosus
B. Foramen ovalis
C. Superficial temporal artery
D. Facial artery
E. Internal carotid artery
F. External carotid artery
G. Vertebral artery
H. Right common carotid artery
I. Right subclavian artery

Color
1. Ductus arteriosus
2. Foramen ovalis
3. Superficial temporal artery
4. Facial artery
5. Internal carotid artery
6. External carotid artery
7. Vertebral artery
8. Right common carotid artery
9. Right subclavian artery

Page 126
Figure 7.8 Arteries serving the brain

Identify
A. Internal carotid artery
B. Basilar artery
C. Vertebral artery

Color
1. Internal carotid artery
2. Basilar artery
3. Vertebral artery

Page 127
Figure 7.9 Arteries of the upper limb

Identify
A. Left common carotid artery
B. Subclavian artery
C. Axillary artery
D. Brachial artery
E. Radial artery
F. Ulnar artery

Color
1. Left common carotid artery
2. Subclavian artery
3. Axillary artery
4. Brachial artery
5. Radial artery
6. Ulnar artery

Page 128
Table 7.1 Anatomical Areas of the Aorta

Identify
A. Ascending aorta
B. Arch of aorta
C. Thoracic aorta
D. Abdominal aorta

Page 129
Figure 7.10 Arteries of the thoracic region

Identify
A. Common carotid arteries
B. Brachiocephalic artery
C. Left subclavian artery
D. Right subclavian artery
E. Posterior intercostal arteries
F. Anterior intercostal arteries
G. Thoracic aorta

Color
1. Brachiocephalic artery
2. Posterior intercostal arteries
3. Anterior intercostal arteries
4. Thoracic aorta

Page 130
Figure 7.11 Branches of the abdominal aorta

Identify
A. Celiac trunk
B. Superior mesenteric artery
C. Renal artery
D. Spermatic/ovarian artery
E. Abdominal aorta
F. Inferior mesenteric artery
G. Common iliac artery

Color
1. Diaphragm
2. Kidney
3. Renal artery
4. Spermatic/ovarian artery
5. Abdominal aorta

Page 131
Figure 7.12 Branches of the celiac artery

Identify
A. Left gastric artery
B. Hepatic artery
C. Splenic artery

Color
1. Celiac artery
2. Left gastric artery
3. Hepatic artery
4. Splenic artery
5. Liver
6. Stomach
7. Spleen

Page 132
Figure 7.13 Branches of the distal
abdominal aorta

Identify
A. Superior mesenteric artery
B. Inferior mesenteric artery
C. Common iliac artery
D. Internal iliac artery
E. External iliac artery

Color
1. Superior mesenteric artery
2. Inferior mesenteric artery
3. Abdominal aorta
4. Common iliac artery

Page 133
Table 7.2 Branches of the Abdominal Aorta
Summarized

Identify
A. Inferior phrenic artery
B. Left gastric artery
C. Hepatic artery
D. Splenic artery
E. Superior mesenteric artery
F. Suprarenal artery
G. Renal artery

H. Ovarian artery
I. Spermatic artery
J. Inferior mesenteric artery
K. Common iliac artery

Page 134
Figure 7.14 Arteries of the pelvis and thigh

Identify
A. Common iliac
B. Internal iliac
C. External iliac
D. Femoral
E. Popliteal

Color
1. Common iliac
2. Internal iliac
3. External iliac
4. Femoral
5. Popliteal

Page 135
Figure 7.15 Arteries of the leg and foot

Identify
A. Popliteal
B. Anterior tibial
C. Peroneal
D. Posterior tibial

Color
1. Popliteal
2. Anterior tibial
3. Peroneal
4. Posterior tibial

Page 136
Figure 7.16 Major systemic veins of
the body

Identify
A. Internal jugular
B. External jugular
C. Brachiocephalic
D. Superior vena cava
E. Cephalic
F. Basilic
G. Hepatic
H. Inferior vena cava
I. Median cubital
J. Greater saphenous

Color
1. Internal jugular
2. External jugular
3. Brachiocephalic
4. Superior vena cava
5. Cephalic
6. Basilic
7. Hepatic
8. Inferior vena cava
9. Median cubital
10. Greater saphenous

Page 137
Figure 7.17 Veins of the head and neck
Figure 7.18 Superficial veins, upper limbs

Identify
A. Left external jugular
B. Left internal jugular
C. Right internal jugular
D. Right brachiocephalic
E. Left brachiocephalic
F. Cephalic
G. Basilic
H. Median cubital
I. Median antebrachial

Color
1. Cephalic
2. Basilic
3. Median cubital
4. Median antebrachial

Page 138
Figure 7.19 Veins of the posterior
thoracic wall

Identify
A. Right brachiocephalic
B. Left brachiocephalic
C. Right subclavian
D. Left internal jugular
E. Left external jugular
F. Right axillary
G. Left subclavian
H. Azygos
I. Hemiazygos
J. Posterior intercostals

Color
1. Azygos
2. Hemiazygos
3. Posterior intercostals

Page 139
Figure 7.20 Veins draining the abdominal
viscera

Identify
A. Hepatic vein
B. Hepatic portal vein
C. Splenic vein
D. Inferior mesenteric vein
E. Superior mesenteric vein

Color
1. Hepatic vein
2. Inferior vena cava
3. Hepatic portal vein
4. Liver
5. Spleen
6. Splenic vein
7. Inferior mesenteric vein
8. Superior mesenteric vein
9. Colon

Page 140
Figure 7.21 Superficial veins of the
lower limb

Identify
A. Femoral (deep)
B. Greater saphenous
C. Popliteal (deep)
D. Lesser saphenous

Color
1. Greater saphenous
2. Lesser saphenous

Page 141
Figure 7.22 The lymphatic system

Identify
A. Entry of thoracic duct into left
   subclavian vein
B. Submandibular lymph nodes
C. Axillary lymph nodes
D. Entry of right lymphatic duct into right
   subclavian vein
E. Thoracic duct
F. Inguinal lymph nodes

Color
1. Area drained by right lymphatic duct
2. Area drained by thoracic duct

MODULE 8:
THE CENTRAL NERVOUS SYSTEM

Page 144
Table 8.1 Divisions and Subdivisions of the
Brain

Identify
A. Cerebrum
B. (Cerebral) cortex
C. Basal ganglia
D. Olfactory bulbs
E. Thalamus
F. Hypothalamus
G. Epithalamus
H. Pineal (body)
I. Corpora quadrigemina
J. Cerebral peduncles
K. Cerebellum
L. Pons
M. Medulla (oblongata)

Page 145
Figure 8.1 (a) Lateral view and (b) frontal cut
of the brain

Identify
A. Longitudinal fissure
B. Central sulcus
C. Lateral sulcus
D. Parieto-occipital sulcus

Color
1. Frontal (lobe)
2. Parietal (lobe)
3. Temporal (lobe)
4. Occipital (lobe)

Definition
a. Primary motor areas, speech, higher
   intellectual thought
b. Vision
c. Primary sensory areas, gustation,
   interpretation of words, shapes, textures
d. Hearing, olfaction, and their interpretation

Page 146
Figure 8.2 Midsagittal section of the brain
Figure 8.3 Frontal section through cerebrum

Identify
A. Third ventricle
B. Corpus callosum
C. Hypothalamus
D. Pituitary gland
E. Lateral ventricle
F. Basal ganglia
G. Thalamus
H. Massa intermedia

Color
1. Corpus callosum
2. Pituitary gland

Page 147
Figure 8.4 Inferior view of the brain
Figure 8.5 Internal impression of the
ventricles

Identify
A. Pituitary gland
B. Cerebral peduncles
C. Pons
D. Cerebellum
E. Medulla
F. Interventricular foramen
G. Cerebral aqueduct
H. Lateral aperture
I. Median aperture

Color
1. Third ventricle
2. Lateral ventricles
3. Fourth ventricle

Page 148
Figure 8.6 Meninges of the brain, frontal section
Figure 8.7 Extensions of the dura within the cranium

Identify
A. Superior sagittal sinus
B. Arachnoid villi
C. Falx cerebri (formed by dura mater)
D. Tentorium
E. Falx cerebelli

Color
1. Dura mater
2. Arachnoid
3. Pia mater
4. Subarachnoid space

Page 149
Figure 8.8 Spinal cord in (a) dorsal and (b) lateral views

Identify
A. Conus medularis
B. Cauda equina

Color
1. Spinal cord
2. Dura mater

Page 150
Figure 8.9 Spinal cord and vertebra, transverse view
Figure 8.10 Structure of the spinal cord and meninges

Identify
A. Dorsal root
B. Dorsal root gangion
C. Ventral root
D. Posterior median sulcus
E. Central canal
F. Anterior median sulcus

Color
1. Pia mater
2. Arachnoid
3. Dura mater

Page 151
Table 8.2 Summary of Spinal White Matter Tracts

Definition
a. P
b. M
c. P
d. M
e. L
f. E
g. A
h. E
i. L
j. N
k. L
l. M,N
m. L
n. M
o. A
p. N
q. L
r. M (Above)
s. A
t. N
u. A
v. M (Above)
w. A
x. M

MODULE 9: THE PERIPHERAL NERVOUS SYSTEM

Page 154
Figure 9.1 Basal view of the brain showing cranial nerve origins

Identify
A. Olfactory I
B. Optic II
C. Oculomotor III
D. Trochlear IV
E. Trigeminal V
F. Facial VII
G. Abducens VI
H. Vestibulocochlear VIII
I. Glossopharyngeal IX
J. Vagus X
K. Hypoglossal XII
L. (Spino)accessory XI

Color
1. Olfactory I
2. Optic II
3. Oculomotor III
4. Trochlear IV
5. Trigeminal V
6. Facial VII
7. Abducens VI
8. Vestibulocochlear VIII
9. Glossopharyngeal IX
10. Vagus X
11. Hypoglossal XII
12. (Spino)accessory XI

Page 155
Figure 9.2 Brain, visual pathway
Figure 9.3 Anterior head, partial innervation

Identify
A. Optic II
B. Optic chiasma
C. Facial VII
D. Sensory fibers (taste)
E. Somatic motor branch

Color
1. Optic II
2. Optic chiasma
3. Facial VII
4. Sensory fibers (taste)
5. Somatic motor branch

Page 156
Figure 9.4 Tongue musculature innervation
Figure 9.5 Eye musculature, partial innervation

Identify
A. Hypoglossal canal
B. Hypoglossal XII
C. Lateral rectus muscle
D. Abducens V

Color
1. Hypoglossal canal
2. Hypoglossal XII
3. Lateral rectus muscle
4. Abducens V

Page 157
Figure 9.6 Nasal mucosa, partial innervation
Figure 9.7 Neck musculature, partial innervation

Identify
A. Olfactory tract
B. Olfactory bulb
C. Olfactory I

D. Cranial root
E. Spinal root
F. (Spino)Accessory XI

Color
1. Olfactory tract
2. Olfactory bulb
3. Olfactory I
4. Cranial root
5. Spinal root
6. (Spino)Accessory XI

Page 158
Figure 9.8 Major viscera, partial innervation
Figure 9.9 Inner ear innervation

Identify
A. Vagus X
B. Vestibular branch
C. Vestibulocochlear VII
D. Cochlear branch

Color
1. Vagus X
2. Vestibular branch
3. Vestibulocochlear VII
4. Cochlear branch

Page 159
Figure 9.10 Tongue and pharynx, partial innervation
Figure 9.11 Eye muscles, partial innervation

Identify
A. Parotid gland
B. Glossopharyngeal IX
C. Sensory fibers (taste)
D. Extrinsic eye muscles
E. Oculomotor III

Color
1. Parotid gland
2. Glossopharyngeal IX
3. Sensory fibers (taste)
4. Extrinsic eye muscles
5. Oculomotor III

Page 160
Figure 9.12 Extrinsic eye muscles, partial innervation
Figure 9.13 Anterior head, partial innervation

Identify
A. Superior oblique muscle
B. Trochlea
C. Trochlear IV
D. Ophthalmic branch
E. Maxillary branch
F. Trigeminal V
G. Mandibular branch
H. Muscles of mastication

Color
1. Superior oblique muscle
2. Trochlea
3. Trochlear IV
4. Ophthalmic branch
5. Maxillary branch
6. Trigeminal V
7. Mandibular branch
8. Muscles of mastication

Page 161
Table 9.1 Cranial Nerve Names and General Functions

Identify
A. Olfactory
B. Figure 9.6

C. Optic
D. Figure 9.2
E. Oculomotor
F. Figure 9.11
G. Trochlear
H. Figure 9.12
I. Trigeminal
J. Figure 9.13
K. Abducens
L. Figure 9.5
M. Facial
N. Figure 9.3
O. Vestibulocochlear
P. Figure 9.9
Q. Glossopharyngeal
R. Figure 9.10
S. Vagus
T. Figure 9.8
U. (Spino)Accessory
V. Figure 9.7
W. Hypoglossal
X. Figure 9.4

Page 162
Figure 9.14 Spinal cord segment and adjacent peripheral nervous system structures
Figure 9.15 Thorax in transverse section, showing formation of intercostal nerves

Identify
A. Dorsal root
B. Dorsal root ganglion
C. Ventral root
D. Spinal nerve
E. Sympathetic chain ganglion
F. Dorsal ramus
G. Ventral ramus
H. Intercostal nerve

Color
1. Dorsal root
2. Dorsal root ganglion
3. Ventral root
4. Spinal nerve
5. Sympathetic chain ganglion
6. Dorsal ramus
7. Ventral ramus
8. Intercostal nerve

Page 163
Table 9.2 Functional Components of Spinal Nerves

Identify
A. Motor
B. Autonomic
C. Parasympathetic
D. Sympathetic
E. Somatic
F. Sensory

Page 164
Figure 9.16 Plexus formation from the ventral ramus of spinal nerves

Identify
A. Cervical plexus
B. Brachial plexus
C. Intercostal nerves
D. Lumbar plexus
E. Sacral plexus

Page 165
Table 9.3 Naming of Peripheral Nervous System Structures

Identify
A. Cranial nerves
B. Spinal nerves
C. Cervical nerves

D. Thoracic nerves
E. Lumbar nerves
F. Sacral nerves
G. Coccygeal nerves

Page 166
Figure 9.17 The cervical plexus
Figure 9.18 The brachial plexus

Identify
A. Phrenic nerve
B. Axillary nerve
C. Radial nerve
D. Musculocutaneous nerve
E. Median nerve
F. Ulnar nerve

Color
1. Phrenic nerve
2. Axillary nerve
3. Radial nerve
4. Musculocutaneous nerve
5. Median nerve
6. Ulnar nerve

Page 167
Figure 9.19 The lumbosacral plexus

Identify
A. Femoral nerve
B. Common peroneal nerve
C. Tibial nerve
D. Sciatic nerve

Color
1. Femoral nerve
2. Common peroneal nerve
3. Tibial nerve
4. Sciatic nerve

MODULE 10: THE SPECIAL SENSES

Page 170
Figure 10.1 External eye muscles in (a) lateral and (b) superior views
Figure 10.2 Oblique view of the eye, including retinal detail

Identify
A. Optic disc
B. Ciliary body
C. Fovea centralis
D. Cornea
E. Macula lutea
F. Pupil
G. Sclera

Color
1. Superior rectus
2. Superior oblique
3. Medial rectus
4. Lateral rectus
5. Inferior oblique
6. Inferior rectus
7. Retina
8. Iris
9. Lens
10. Choroid

Page 171
Figure 10.3 Ciliary body and adjacent structures

Identify
A. Anterior chamber
B. Posterior chamber
C. Suspensory ligaments
D. Canal of Schlemm
E. Sclera

Color
1. Conjunctiva
2. Iris
3. Ciliary muscles
4. Choroid

Page 172
Figure 10.4 Anatomical components of the ear
Figure 10.5 Membranous labyrinth suspended in the osseous labyrinth, greatly enlarged

Identify
A. Auricle (pinna)
B. Oval window
C. Round window
D. External auditory meatus
E. Utricle
F. Saccule
G. Ampulla

Color
1. Tympanic membrane
2. Auditory (Eustachian) tube
3. Stapes
4. Incus
5. Malleus
6. Perilymph
7. Endolymph

Page 173
Figure 10.6 Cross section through one turn of the cochlea
Figure 10.7 Enlarged section through the spiral organ

Identify
A. Scala vestibuli
B. Tectorial membrane
C. Scala media (cochlear duct)
D. Cochlear nerve (subpart of the vestibulocochlear nerve)
E. Basilar membrane
F. Scala tympani

Color
1. Perilymph
2. Endolymph
3. Cochlear nerve (subpart of the vestibulocochlear nerve)

**MODULE 11:
THE ENDOCRINE SYSTEM**

Page 176
Figure 11.1 The pituitary (hypophysis)
Figure 11.2 Hypothalamus and pituitary

Identify
A. Neurohypophysis
B. Adenohypophysis
C. Growth hormone
D. Thyrotropin
E. Adrenocorticotropin
F. Follicle stimulating hormone
G. Luteinizing hormone
H. Prolactin
I. Antidiuretic hormone
J. Oxytocin

Color
1. Hypothalamus
2. Neurohypophysis
3. Adenohypophysis

Page 177
Figure 11.3 Midsagittal section through brain
Figure 11.4 Anterior view at larynx

Identify
A. Pineal gland
B. Melatonin
C. Thyroid gland
D. Thyroxin
E. (Thyro)calcitonin

Color
1. Third ventricle
2. Pineal gland
3. Larynx
4. Thyroid gland

Page 178
Figure 11.5 Posterior view at larynx
Figure 11.6 Pancreas

Identify
A. Superior parathyroid glands
B. Inferior parathyroid glands
C. Parathyroid hormone
D. Alpha cells
E. Islets of Langerhans
F. Beta cells
G. Glucagon
H. Insulin

Color
1. Thyroid gland
2. Pancreas
3. Alpha cells
4. Beta cells

Page 179
Figure 11.7 Posterior abdominal wall, anterior view
Figure 11.8 Adrenal glands

Identify
A. Adrenal glands
B. Cortex
C. Medulla
D. Zona glomerulosa
E. Mineralocorticoids
F. Zona fasciculata
G. Glucocorticoids
H. Zona reticularis
I. Androgens
J. Estrogens
K. Epinephrine
L. Norepinephrine

Color
1. Adrenal glands
2. Kidney
3. Cortex
4. Medulla
5. Zona glomerulosa
6. Zona fasciculata
7. Zona reticularis

Page 180
Figure 11.9 Ovary, greatly enlarged
Figure 11.10 Testes, microscopic section

Identify
A. Corpus luteum
B. Graafian follicle
C. Developing follicles
D. Estrogen
E. Progesterone
F. Interstitial cells
G. Testosterone

Color
1. Corpus luteum
2. Graafian follicle
3. Ovary
4. Developing follicles
5. Interstitial cells

6. Testis

Page 181
Figure 11.11 Relative locations of the endocrine glands

Identify
A. Pineal
B. Pituitary (hypophysis)
C. Parathyroid
D. Thyroid
E. Thymus
F. Adrenal
G. Pancreas
H. Ovary
I. Testes

Color
1. Pineal
2. Pituitary
3. Parathyroid
4. Thyroid
5. Thymus
6. Adrenal
7. Pancreas
8. Ovary
9. Testes

MODULE 12: THE RESPIRATORY SYSTEM

Page 184
Figure 12.1 Sagittal section through head and neck

Identify
A. Superior meatus
B. Internal nares
C. Middle meatus
D. Auditory (Eustachian) tube
E. Inferior meatus
F. External nares

Color
1. Superior nasal concha
2. Middle nasal concha
3. Inferior nasal concha
4. Hard palate
5. Soft palate
6. Pharyngeal tonsil (adenoids)
7. Lingual tonsil
8. Palatine tonsil
9. Epiglottis
10. Thyroid cartilage
11. Cricoid cartilage

Page 185
Figure 12.2 The larynx in (a) anterior view, and (b) sagittal section
Figure 12.3 Superior view of larynx with (a) glottis closed and (b) glottis open

Identify
A. Ventricular fold (false vocal cord or vestibular fold)
B. True vocal cord (vocal fold)
C. Glottis

Color
1. Epiglottis
2. Ventricular fold (vestibular fold)
3. Thyroid cartilage
4. True vocal cord
5. Cricoid cartilage
6. Arytenoid cartilage(s)

Page 186
Figure 12.4 Structures of the lower respiratory system: (a) lungs, trachea, and bronchi

(b) microscopic enlargement

Identify
A. Trachea
B. Primary bronchi
C. Secondary (lobar) bronchi
D. Tertiary bronchi
E. Terminal bronchiole
F. Respiratory bronchioles
G. Alveoli

Color
1. Superior lobes
2. Middle lobe
3. Inferior lobes

MODULE 13: THE DIGESTIVE SYSTEM

Page 188
Figure 13.1 Structures of the gastrointestinal tract

Identify
A. Oral cavity
B. Pharynx
C. Esophagus
D. Stomach
E. Duodenum
F. Transverse colon
G. Descending colon
H. Ascending colon
I. Ileum
J. Cecum
K. Rectum
L. Jejunum

Color
1. Oral cavity
2. Pharynx
3. Esophagus
4. Stomach
5. Duodenum
6. Transverse colon
7. Descending colon
8. Ascending colon
9. Ileum
10. Cecum
11. Rectum
12. Jejunum

Page 189
Figure 13.2 Anterior-inferior head and neck, sagittal section
Figure 13.3 Permanent teeth

Identify
A. Oral cavity
B. Uvula
C. Lingual frenulum
D. Oropharynx
E. Laryngopharynx
F. Incisors
G. Canines
H. Premolars
I. Molars

Color
1. Oral cavity
2. Lingual frenulum
3. Uvula
4. Oropharynx
5. Laryngopharynx
6. Incisors
7. Canines
8. Premolars
9. Molars

Page 190
Figure 13.4 Molar (tricuspid), vertical section
Figure 13.5 The major salivary glands

Identify
A. Crown
B. Neck
C. Root
D. Parotid gland
E. Submandibular gland
F. Sublingual gland

Color
1. Crown
2. Neck
3. Root
4. Parotid gland
5. Submandibular gland
6. Sublingual gland

Page 191
Figure 13.6 Stomach and duodenum

Identify
A. Cardiac orifice
B. Fundus
C. Lesser curvature
D. Body
E. Duodenum
F. Pyloric orifice
G. Pylorus
H. Greater curvature
I. Rugae
J. Plicae circulares
K. Villi

Color
1. Esophagus
2. Fundus
3. Body
4. Duodenum
5. Pylorus
6. Pancreas
7. Villi

Page 192
Figure 13.7 The liver
Figure 13.8 Exocrine ducts entering the duodenum

Identify
A. Coronary ligament
B. Falciform ligament
C. Round ligament
D. Hepatic duct
E. Cystic duct
F. Common bile duct
G. Gallbladder
H. Hepatopancreatic ampulla
I. Duodenal papilla
J. Pancreatic duct
K. Pancreas

Color
1. Right lobe
2. Left lobe
3. Gallbladder
4. Caudate lobe
5. Hepatic duct
6. Cystic duct
7. Common bile duct
8. Pancreatic duct
9. Pancreas

Page 193
Figure 13.9 The large intestine

Identify
A. Taenia coli
B. Epiploic appendages

C. Haustra
D. Ileocecal valve
E. Cecum
F. Appendix
G. Rectum
H. Anal sphincter (external)

Color
1. Transverse colon
2. Ascending colon
3. Descending colon
4. Taenia coli
5. Cecum
6. Sigmoid colon
7. Appendix
8. Rectum
9. Anal sphincter

Page 194
Figure 13.10 Peritoneal attachments, sagittal section

Identify
A. Coronary ligament
B. Visceral peritoneum
C. Parietal peritoneum
D. Lesser omentum
E. Mesocolon
F. Greater omentum
G. Intestinal mesenteries

Color
1. Coronary ligament
2. Visceral peritoneum
3. Parietal peritoneum
4. Lesser omentum
5. Mesocolon
6. Greater omentum
7. Intestinal mesenteries

Page 195
Figure 13.11 Abdominal viscera, superficial view

Identify
A. Falciform ligament
B. Greater omentum

Color
1. Falciform ligament
2. Liver
3. Stomach
4. Gallbladder
5. Greater omentum
6. Cecum
7. Large intestine
8. Small intestine

MODULE 14:
THE URINARY SYSTEM

Page 198
Figure 14.1 The urinary system, anterior view

Color
1. Kidneys
2. Ureters
3. Urinary bladder
4. Urethra

Page 199
Figure 14.2 Transverse section through the trunk at L₁
Figure 14.3 Longitudinal section through a kidney

Identify
A. Renal hilus

B. Interlobar vessels
C. Arcuate vessels
D. Interlobular vessels
E. Renal pelvis
F. Major calyx
G. Renal pyramid (within medulla)
H. Minor calyx

Color
1. Peritoneum
2. Perirenal fat
3. Renal fascia
4. Renal capsule
5. Cortex
6. Medulla

Page 200
Figure 14.4 Detail of a nephron and its blood supply (400x)

Identify
A. Peritubular capillaries
B. Glomerulus
C. Efferent arteriole
D. Afferent arteriole
E. Interlobular artery
F. Arcuate artery
G. Arcuate vein
H. Interlobular vein

Color
1. Proximal convoluted tubule
2. Bowman's (glomerular) capsule
3. Distal convoluted tubule
4. Collecting duct
5. Loop of Henle: ascending limb
6. Loop of Henle: descending limb

Page 201
Figure 14.5 (a) Frontal section of female urinary bladder, and (b) sagittal section of male pelvis and urinary bladder

Identify
A. Orifice of ureter
B. Trigone
C. Urethra

Color
1. Internal sphincter
2. External sphincter

Page 202
Table 14.1 Summary of Excretory Pathway Through the Urinary System

Identify
A. Interlobar arteries
B. Arcuate arteries
C. Interlobular arteries
D. Afferent arterioles
E. Glomerulus
F. Bowman's (glomerular) capsule
G. Proximal convoluted tubule
H. Loop of Henle
I. Distal convoluted tubule
J. Collecting duct
K. Minor calyce
L. Major calyces
M. Renal pelvis
N. Ureter
O. Urethra

MODULE 15: THE REPRODUCTIVE SYSTEM

Page 204
Figure 15.1 Human testis, sectioned
Figure 15.2 Male pelvis, sagittal section

Identify
A. Spermatic cord
B. Vas deferens
C. Epididymis
D. Seminiferous tubules
E. Ampulla
F. Inguinal canal
G. Ejaculatory duct
H. Membranous urethra
I. Penis
J. Cavernous urethra
K. Scrotum
L. Prepuce

Color
1. Spermatic cord
2. Vas deferens
3. Epididymis
4. Seminiferous tubules
5. Ampulla
6. Penis
7. Testes
8. Scrotum

Page 205
Figure 15.3 Male genitalia

Identify
A. Seminal vesicle
B. Prostate
C. Bulbourethral gland
D. Corpus cavernosum
E. Bulb
F. Crus
G. Corpus spongiosum
H. Glans

Color
1. Ampulla
2. Seminal vesicle
3. Vas deferens
4. Prostate
5. Bulbourethral gland
6. Bulb
7. Crus
8. Corpus cavernosum

9. Corpus spongiosum
10. Epididymis
11. Testes
12. Glans

Page 206
Figure 15.4 Human ovary, greatly enlarged
Figure 15.5 Female reproductive structures

Identify
A. Corpus luteum
B. Graafian follicle
C. Ovary
D. Ostium
E. Mesometrium
F. Oviduct
G. Fundus
H. Fimbria
I. Body
J. Endometrium
K. Myometrium
L. Cervix
M. Vagina

Color
1. Corpus luteum
2. Graafian follicle
3. Ovary
4. Ostium
5. Oviduct
6. Fundus
7. Fimbria
8. Mesometrium
9. Body
10. Endometrium
11. Myometrium
12. Vagina
13. Cervix

Page 207
Figure 15.6 Female reproductive structures, supportive elements
Figure 15.7 Female external genitalia

Identify
A. Suspensory ligament

B. Mesosalpinx
C. Broad ligament
D. Mesovarium
E. Mesometrium
F. Ovarian ligament
G. Round ligament
H. Mons pubis
I. Labia majora
J. Clitoris
K. Prepuce
L. Labia minora
M. Vestibule
N. Hymen
O. Perineum
P. Vaginal orifice

Color
1. Suspensory ligament
2. Mesosalpinx
3. Mesovarium
4. Mesometrium
5. Ovarian ligament
6. Round ligament
7. Mons pubis
8. Labia majora
9. Prepuce
10. Labia minora
11. Perineum

Page 208
Figure 15.8 Female pelvis, sagittal section

Color
1. Suspensory ligament
2. Fimbria
3. Ovary
4. Oviduct
5. Uterus
6. Broad ligament
7. Round ligament
8. Cervix
9. Mons pubis
10. Vagina
11. Clitoris
12. Labia minora
13. Labia majora